Microwave-Assisted Concrete Technology

Production, Demolition and Recycling

Microwave-Assisted Concrete Technology

Production, Demolition and Recycling

K.C. Gary Ong
National University of Singapore

Ali Akbarnezhad
University of New South Wales

CRC Press
Taylor & Francis Group
Boca Raton London New York

CRC Press is an imprint of the
Taylor & Francis Group, an **informa** business

A SPON PRESS BOOK

CRC Press
Taylor & Francis Group
6000 Broken Sound Parkway NW, Suite 300
Boca Raton, FL 33487-2742

First issued in paperback 2017

© 2015 by Taylor & Francis Group, LLC
CRC Press is an imprint of Taylor & Francis Group, an Informa business

No claim to original U.S. Government works

ISBN-13: 978-1-4665-8393-1 (hbk)
ISBN-13: 978-1-138-74889-7 (pbk)

Library of Congress Cataloging-in-Publication Data

Ong, K. C. G.
   Microwave-assisted concrete technology : production, demolition, and recycling / author/editor, Gary Ong and Ali Akbarnezhad.
      pages cm
   Includes bibliographical references and index.
   ISBN 978-1-4665-8393-1
   1. Concrete--Heating. 2. Concrete--Effect of temperature on. 3. Microwave heating.
I. Akbarnezhad, Ali. II. Title.

TP882.O54 2015
666'.893--dc23   2014027987

Visit the Taylor & Francis Web site at
http://www.taylorandfrancis.com

and the CRC Press Web site at
http://www.crcpress.com

K. C. G. Ong dedicates this book to Patricia, James, Pamela and Mom for their love and support and in memory of Pa.

Ali Akbarnezhad would like to dedicate this book to his wife, Sara, for her love and support.

Contents

Preface

The incredible advances in science and technology over the last century have led to the discovery of new forms of energy and efficient energy transfer technologies that have been implemented recently in material-processing applications in many industries. One such example is microwave heating, discovered by Percy Spencer in 1945. Because of its higher efficiency, less-polluting nature, and unique capabilities in adjusting the pattern of heating, microwave heating is rapidly replacing conventional heating technologies used in many industries, including food processing, timber drying, ceramic fabrication and processing, mining, and so on. However, despite the widespread use in such industries, technologies based on microwave heating are rarely adopted in the construction industry. This is partially because of the lack of awareness of civil engineering professionals about the improved productivity achievable in using microwave-assisted technologies.

One potential area of application for microwave-assisted technologies in civil engineering is construction material processing. Concrete, the most commonly used construction material in civil engineering, has great potential to take advantage of microwave-assisted technologies at the various stages within its life cycle. This is mainly because of the dielectric nature of concrete, rendering it responsive to an electromagnetic field. Microwave heating has been recently proposed for a number of innovative applications in the concrete industry. Examples of actual applications in recent years include microwave curing of prefabricated concrete elements, microwave-assisted demolition of concrete structures, microwave drilling and cutting of concrete, and microwave-assisted concrete recycling. The advantages of replacing the conventional concrete curing, demolition, and recycling techniques with the more efficient microwave-assisted techniques are unequivocal and have been demonstrated through a number of experimental and analytical studies reported in available literature. However, stakeholders in the generally conservative construction industry are slow in embracing it. Microwave technology is also not part of any traditional civil engineering curriculum, which is compounded by the increasing pace of specialisation in modern practice in the various fields of engineering today. Paradoxically, today's civil engineers are expected to be more innovative and have knowledge of sustainability issues, which invariably calls for dissolving the boundaries between technological, scientific, humanistic, and social issues.

With this in mind, the main objective of this book is to reduce the psychological barriers to the adoption of microwave-assisted technologies in civil engineering through providing civil engineering professionals with a basic understanding of microwave-assisted technologies and their applications in the concrete industry. We hope that further adoption of microwave-assisted concrete technologies by the concrete industry and further research into this area can be encouraged by highlighting the advantages achievable through adoption of these technologies and elimination of the misconceptions about disadvantages and difficulties in implementation.

To the best of our knowledge, this book is the first reference book dealing with the underlying concepts of microwave heating of concrete from the perspective of practitioners of the construction industry, especially the modern concrete industry and the impact sustainable development has at the technological and environmental level. The book is put forward as a platform for the collection of relevant state-of-the-art technologies covering the design of the industrial microwave heating generators and applicators, as well as the various process control techniques available to monitor the condition of concrete during the heating process. The book aims to address the gap in civil engineering education as a singular reference to the various techniques in the use of microwave heating in the modern concrete industry. This book could be of help to stakeholders in the concrete industry, including civil engineers, concrete technology experts and practitioners, and researchers, through describing the various actual and potential applications, the underlying phenomena, the design concepts of microwave heating systems, and the various process control techniques available for achieving efficiency.

After covering the basics of concrete-microwave field interactions in Chapter 1, a number of innovative microwave-assisted concrete technologies for use in the production, demolition, and recycling of concrete as well as the control mechanisms required to ensure the efficiency of such methods are introduced in Chapters 2 to 5. Finally, Chapter 6 presents a brief introduction to the design of microwave heating applicators suitable for use with the applications discussed in Chapters 2 to 4. A selected bibliography is provided at the end of each chapter to guide interested readers to further information.

Finally, we would like to express our interest in establishing contact with the professionals in the fields of concrete and microwave technology and readers of this book to further improve future editions of this book.

We can be reached via e-mail: ceeongkc@nus.edu.sg (K.C.G. Ong) and a.akbarnezhad@unsw.edu.au (Ali Akbarnezhad).

K.C.G. Ong
Associate Professor
National University of Singapore

Ali Akbarnezhad
Lecturer
The University of New South Wales

About the Authors

Dr K.C. Gary Ong is an associate professor and deputy head (administration) in the Department of Civil and Environmental Engineering, National University of Singapore. He is an alumnus of the University of Singapore and the University of Dundee (UK). He is a past president of the American Concrete Institute (ACI), Singapore Chapter, and the Singapore Concrete Institute. Professor Ong received the inaugural IES, Engineer KYD Gin Award for Prestigious Publication (Application) in 1990, the IES/IStructE Best Paper Award in 1993, the Japan Concrete Institute's JCI-OWICS 2006 Award in August 2006, and the Singapore Concrete Institute Excellence Award 2013 (Innovators Category) in November 2013. He is a member of the editorial boards of the ASTM *Journal of Testing and Evaluation* and the *Journal of Ferrocement*. He is a member of the International Organization for Standardisation (ISO) Technical Committee, ISO/TC71/SC4, Performance Requirements for Structural Concrete, and the ISO Technical Committee, ISO/TC71/SC5, Simplified Design Standard for Concrete Structures. He is a member of the Building and Construction Standards Committee (SPRING Singapore) and is convenor of WG2 for the Migration to Eurocodes in Singapore. He received a Merit Award from the Singapore Standards Council in 1993. Professor Ong has been a technical assessor for the Singapore Laboratory Accreditation Scheme (SINGLAS/ISO) since 1991, for which he was recognized with a Bronze Award in 1995, a Silver Award in 1998, and a Gold Award in 2005. He received the SAC Merit Award from the Singapore Accreditation Council in 2008 and the SPRING Singapore Merit Award in 2014. He is presently chairman of the Technical Committee for Civil Engineering Testing. His research interests include advanced composite materials, microwave-assisted concrete technology, durability and assessment of concrete structures, retrofitting, and sustainability; he has edited/coedited 15 conference proceedings and published more than 200 publications in journals and conference proceedings.

Dr Ali Akbarnezhad is a lecturer in the School of Civil and Environmental Engineering, University of New South Wales. His main areas of research include concrete technology, design for sustainability, sustainability assessment, construction technology, and information technology; he has published more than 30 papers in leading refereed journals and conference proceedings. He is an alumnus of the Amirkabir University of Technology (Tehran) and National University of Singapore. Dr Akbarnezhad has been involved in microwave heating research and development for more than 8 years and is the inventor of two microwave-assisted techniques: microwave-assisted beneficiation of recycled concrete aggregates and microwave-assisted removal of tiles and ceramic finishes from concrete floors and walls. He was the recipient of the Japan Concrete Institute's JCI-OWICS Award in 2006 and the Skilled Builders Award from the Building and Construction Authority of Singapore in 2012.

Chapter 1

Introduction

1.1 MICROWAVE HEATING

Microwaves are a portion of the electromagnetic (EM) spectrum lying between UHF (ultrahigh-frequency) radio waves and heat (infrared) waves and span the frequency range of 300 MHz to 300 GHz (Figure 1.1). Since the 1950s, microwaves have been widely used in a variety of applications, including communication, radio detection and ranging (radar), radio astronomy, navigation, spectroscopy, and heating and power applications.

Discovering the heating capabilities of microwaves was an accidental by-product of investigations on radar technology. Intensive research during World War II into high-definition microwave radar transmitters led to the development of generators of microwave of various frequencies, in particular the magnetron valve as microwave generators. Percy Spencer, an American engineer, was working on an active radar set when he noticed that the candy bar in his pocked had melted. He then tried to verify his discovery by placing some popcorn kernels near the tube and watched amazed as the popcorn sputtered, cracked, and popped all over his lab. Spencer then created the first microwave oven (as known today) by attaching a high-density microwave generator to an enclosed metal box, allowing for controllable and safe experimentation. The first patent for a cooking microwave oven was filled in October 1945 by the Raytheon company, Spencer's employer (Figure 1.2). The first commercial microwave oven, named the Radarange, weighed about 750 pounds and was about 6 feet tall (Figure 1.3) [1]. The research and development for applications of microwave heating and improving its efficiency have been intensive since its discovery and continue to this day. The exceptional efficiency of the microwave generators (e.g., ~85% at 900 MHz and ~80% at 2450 GHz) has turned microwave heating into a viable potential replacement for the polluting and high-energy-consuming conventional heating [2].

The largest market for microwave heating is the domestic microwave oven. However, commercial microwave heating units are also widely used in the food industry for tempering, thawing, and continuous baking as well

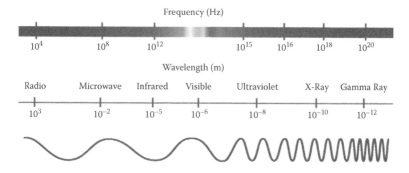

Figure 1.1 Electromagnetic spectrum.

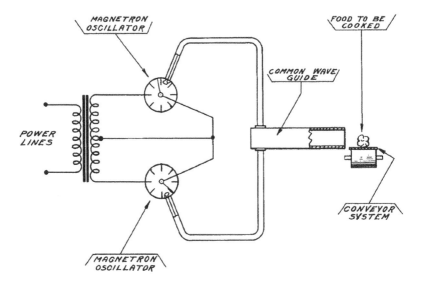

Figure 1.2 Configuration of the first microwave heating system patented by Spencer et al. (1945).

as in many other industries, including the vacuum drying, pasteurisation and sterilisation processing common in the ceramics, rubber, and plastic industries [3]. Figure 1.4 shows a large industrial microwave tunnel used for drying purposes.

To minimise the interference with the microwave frequencies used in communication applications, specific frequencies, often known as ISM (industrial, scientific, and medical) frequencies, have been allocated for microwave heating applications [3]. In general, communications equipment should not be affected by any interference generated by ISM equipment. Among the ISM frequencies, 915 MHz, 2.450 GHz, 10.6 GHz, and 18 GHz

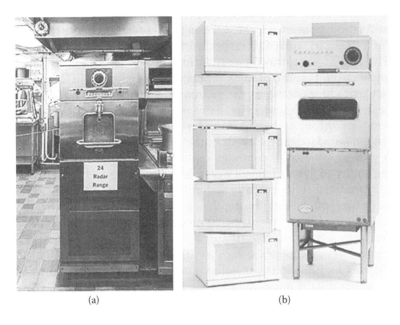

(a)   (b)

Figure 1.3 (a) Raytheon Radarange aboard the *NS Savannah* nuclear-powered cargo ship, installed circa 1961. (b) Radarange compared to typical modern microwave ovens.

Figure 1.4 An industrial microwave tunnel dryer.

are the most commonly used frequencies for microwave heating applications. At these frequencies, the wavelength of radiation in the transmitting medium (dielectric material) ranges from 1 to 10 cm. The most commonly used source of microwave energy, primarily for reasons of efficiency, is the magnetron. Because of mass production, magnetrons at 2.45 GHz are particularly cheap; however, magnetrons are available for other frequency ranges as well. Other sources available include traveling wave tube (TWT), klystron, gyrotron, and solid-state devices. Each has characteristics that can be exploited to suit the needs of the user.

At present, almost all common applications of microwave heating rely on the ability of lower-frequency (916 MHz and 2.45 GHz) microwaves to heat the material volumetrically and uniformly. This is because, at this frequency range, the penetration depth of microwaves usually exceeds the typical thickness of the materials being processed by most current microwave heating user industries, resulting in a uniformly heated material. Higher ISM microwave frequencies, such as 10.6 and 18 GHz, usually dissipate much faster in the dielectric materials and thus may result in localized heating, which is usually undesired for cooking and drying purposes. However, although not suitable for cooking and drying purposes, higher microwave frequencies have been shown to be highly effective in applications for which localised and selective heating of materials is crucial. These include the applications in cutting, drilling, and demolition of hard dielectric materials such as hard rock and concrete, for which localised microwave heating is used to generate high-differential thermal stresses. The use of microwave heating in such an application can significantly improve the efficiency and result in considerable economic and environmental benefits. With recent technological developments in microwave technology, especially in producing cheaper high-frequency cum high-power microwave generators, the use of microwave heating technology in such innovative applications in place of the less-efficient traditional heating technologies makes more sense than ever before.

1.2 APPLICATIONS OF MICROWAVE HEATING IN CONCRETE TECHNOLOGY

The unique capabilities of microwave heating have also attracted the attention of civil engineers, especially concrete specialists. Since the 1980s, microwave heating has found various applications in concrete production, demolition, and recycling. These applications take advantage of the capabilities of a range of microwave frequencies to heat concrete to different extents and degrees of uniformity. Replacement of the often-inefficient conventional techniques with microwave-assisted techniques may significantly decrease the acknowledged negative environmental impacts associated with concrete during its life cycle. In this book, the applications of microwave heating in concrete technology are categorised into three areas as follows:

1. Microwave-assisted accelerated curing of concrete
2. Microwave-assisted selective demolition and drilling of concrete
3. Microwave-assisted recycling of concrete

These applications are briefly explained in the sections that follow.

1.2.1 Microwave-assisted accelerated curing of concrete

Precast concrete is widely used in modern construction to increase the speed of construction and to enhance the quality of the concrete products. Precast concrete components are normally fabricated in the prefabrication plant and then shipped to the construction site for installation [4]. An important aspect affecting the production rate of a prefabrication plant and thus the speed of construction is the time required for concrete to gain enough strength to allow for demoulding and transportation of components to the construction site. At present, the use of special cement compositions or conventional curing at elevated temperatures (with or without steam) is commonly used to accelerate the strength development of the precast concrete components. The use of such techniques, however, has its drawbacks and limitations. For instance, conventional curing at elevated temperatures typically takes more than 24 hours for concrete to reach sufficient strength. Such a long heating duration leads to significant energy use, carbon footprint, and costs. Moreover, the use of chemical accelerators and rapid-hardening cements may cause long-term durability problems that are as yet not well understood [5].

Microwave curing is believed to have great potential to revolutionize the curing of precast concrete [5]. Microwave heating at lower ISM frequencies (such as 915 and 2450 MHz) can be used to heat the concrete uniformly and significantly reduce time required for curing. Previous studies have shown that, unlike steam-cured concrete, microwave-cured concrete can gain considerable strength in just a few hours without compromising its long-term properties [5–7]. Such reduced curing times can significantly decrease energy consumption and thus the carbon emissions associated with conventional curing processes aside from the other advantages arising from less energy consumed for storage of the precast components before being transported for installation. Chapter 2 of this book reviews the recent achievements in using microwave curing for accelerating the strength development rate of freshly cast concrete. A detailed description of the phenomena involved in the microwave curing of concrete as well as the factors affecting curing efficiency is presented.

1.2.2 Microwave-assisted selective demolition and drilling of concrete

Demolition and drilling of concrete using conventional techniques generate considerable noise and a lot of dust. Moreover, a major problem associated with the current state-of-the-art concrete demolition techniques is the lack of selectivity. Because of concrete toughness, it is difficult to demolish or drill into one part of a concrete component without affecting the surrounding parts. A selective concrete demolition technique to precisely remove a specific portion with a large surface area of a concrete component without affecting the surrounding concrete and finishes is in high demand in many repair and retrofitting applications. An example of this is removal of the chloride-contaminated surface of the concrete in roads, offshore and marine structures, and underground concrete structures for replacement with a new concrete layer without adversely affecting the underlying layer. Selective demolition allows reusing the uncontaminated concrete, thereby extending the service life, and has the potential to significantly improve sustainability by preserving the embodied energy already invested in fabricating the original concrete components that are retained.

Another area of application may involve the need to accurately remove a specific part or portion of a concrete component. An example of such applications is in the decontamination of nuclear facilities and facilities used for storage of hazardous materials. Decontamination is an important stage in the decommissioning of old nuclear power plants, nuclear waste-processing plants, and hazardous material storage facilities [8]. For instance, consider the parts of a nuclear power plant or waste-processing plant in which the concrete surface is exposed to radioactive radiation throughout their service life. As a consequence of long-term exposure, various radionuclides diffuse into the concrete, contaminating the surface layer. The thickness of the contaminated layer depends on concrete diffusivity and the exposure duration and is usually between 1 and 10 mm [9]. In conventional methods, contaminated concrete components are demolished in their entirety and disposed of as hazardous wastes, requiring costly precautions and sophisticated disposal techniques. However, as contamination is generally confined to the thin surface layer of the exposed concrete, the use of a selective removal technique to remove only the contaminated surface layer can result in considerable savings in the cost of disposal. The rest of the uncontaminated portions can be disposed of conventionally.

Microwave heating at higher ISM frequencies has been recommended as a potential replacement for conventional concrete surface removal techniques [10,11]. In the microwave decontamination technique, also known as the microwave-assisted concrete removal technique, microwave heating at higher frequencies (>2.45 GHz) and at high power (e.g., >10 kW) is used to heat the thin surface layer of the concrete, thereby generating a region of high thermal stress and pore pressure between the concrete's microwave-exposed surface and its cooler interior. Spalling of the concrete surface occurs when the pore pressure or the thermal stresses generated exceed the concrete's tensile or compressive strengths, respectively. Such microwave-assisted demolition of concrete is highly selective and almost noiseless as it does not involve any mechanical impact.

In another application, a similar concept with minor technical changes has been used recently to develop a novel technique for drilling into concrete components with little noise and dust generation. In Chapter 3, the phenomena associated with the removal of the surface layer of concrete when heated with microwaves as well as the working principles of the microwave drilling technique are described. Moreover, the effects of various factors, including the microwave frequency, microwave power, and water content of concrete, on the efficiency of these methods are discussed.

1.2.3 Microwave-assisted recycling of concrete

Concrete recycling is an increasingly common method for disposing of demolition debris and at the same time providing a sustainable source of concrete aggregates through crushing of the concrete debris and sieving into specific size fractions for use as aggregates in concrete [12,13]. It has been shown that up to 90% of structural concrete debris may be used to produce recycled aggregates of an acceptable quality [14]. Besides serving as an alternative source of aggregate, concrete recycling contributes to reducing landfill spaces needed for the disposal of construction debris. In addition, lower transportation cost and reduced environmental impact are among other advantages of concrete recycling [15].

The use of recycled concrete aggregate (RCA) in construction is a subject of high priority in the building industry throughout the world [16]. However, the RCAs currently produced are usually of low quality and generally are considered unsuitable for use in ready-mix concrete. They are mainly used as base and subbase materials in road carriageway construction or mixed in small fractions, up to 20%, with natural aggregate (NA) to be used in ready-mix concrete. Two main reasons have been suggested as the causes of the lower quality of the RCAs compared to NAs:

1. Because of the incomplete removal of concrete finishes before the demolition of the buildings, there are normally nonconcrete impurities present in the concrete debris that may adversely affect the properties of the RCAs.

2. RCAs normally contain between 20% and 60% mortar, which may be present in the form of entire chunks of mortar or as a layer of mortar adhering to natural coarse aggregates. This mortar portion is normally of a much weaker nature compared to NAs, adversely affecting the mechanical properties of the RCA.

A number of methods have been proposed to increase the quality of RCA through eliminating these causes. However, these methods have a number of significant drawbacks and limitations, rendering them unsuitable for use on an industrial scale. We have recently proposed two microwave-assisted techniques to replace the less-efficient conventional methods used currently. In the first technique, a concept rather similar to that used in microwave-assisted concrete demolition is used to remove the surface finishes from the concrete elements before demolition of the entire building. Incorporating such a removal stage before demolition of a building can significantly reduce the amount of impurities and contaminants present in the concrete debris.

The second technique, termed microwave-assisted mortar separation or microwave beneficiation of RCA, is used to improve the quality of RCA by reducing the amount of mortar present. The microwave-assisted mortar separation technique takes advantage of the differences between the dielectric properties of NAs and mortar, capitalising on the difference in water absorption rates, to develop high-differential thermal stresses in the mortar component, especially at the interface with the aggregates, to break up and separate the mortar from the stone aggregates.

The improved quality of the RCAs after the beneficiation process may significantly encourage the use of RCA as a replacement for NAs, thereby contributing to sustainability by reducing the energy needed for extracting and processing of NAs for use, aside from the energy used for transporting quarried NAs from remote areas to construction sites. The fundamentals and working principles of these techniques are described in Chapter 4. Moreover, the effects of such processes on the properties of the aggregates produced and the concrete made using such aggregates are explained.

1.3 FUNDAMENTALS OF MICROWAVE HEATING

Realising the potential applications of microwave heating in the concrete industry requires an understanding of the fundamentals of electromagnetism, especially in the microwave frequency range. This section aims to provide a basic understanding of microwave properties and microwave behaviour in dielectric materials.

1.3.1 Electromagnetic fields

As mentioned, microwaves comprise a portion of the EM spectrum. Therefore, understanding the behaviour of microwaves requires some basic knowledge of electromagnetism and EM field properties. An EM field is generally a physical field generated by electrically charged objects and is considered one of the four fundamental forces of nature, along with gravitation, weak interaction, and strong interaction. The EM field may extend indefinitely as an EM wave through space and affect the behaviour of the charged objects in the vicinity of the field.

Two fundamental components are necessary for a propagating EM wave to exist: an electric field and a magnetic field. Electric fields are produced by stationary charges; the magnetic fields are generated by moving charges (currents). The electric field in particular, being the prime source of energy transfer to the material being exposed to the microwaves, is a recurring parameter in microwave heating and familiarity with it is essential. In subsequent sections, the nature of electric and magnetic fields and their relationships are briefly discussed.

1.3.1.1 Electric fields

The concept of an electric field was first introduced by Michael Faraday. An electric field is a physical quantity associated with any point surrounding an electrically charged particle or time-varying magnetic field. An electric field exerts a force on other electrically charged objects.

The electric field is a vector field. In SI units, an electric field is characterised by volts per meter (V m^{-1}) or, equivalently, Newton per coulomb (N C^{-1}). The strength (magnitude) of an electric field at a given point in space and time is defined as the magnitude of the force that would be exerted on a positive 1-C charge placed at that point:

$$E = \frac{F_e}{q_t} \tag{1.1}$$

where E is the electric field strength, F_e is the force (N) exerted on the test particle, and q_t is the charge of the particle.

The simplest case for illustrating an electric field in the macroscale may be two conductive plates connected to a voltage V (volts) and placed apart by a distance d (meters), small compared to the size of the plates (Figure 1.5). In this case, an almost-uniform electric field with the strength of V/d volts per meter will be developed between the plates. In this example, while the electric field everywhere between the plates is substantially constant, the geometric discontinuity at the edges causes a local field distortion that increases the field intensity. Ignoring the small field distortion at the edges,

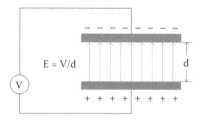

Figure 1.5 The uniform electric field between two conductive plates.

the magnitude of the electric field for this example may be obtained using the following simple equation:

$$E = -\frac{V}{d} \tag{1.2}$$

where V is the voltage difference between the plates, and d is the distance separating the plates. It is also immediately obvious that the value of the electric field tangential to the plates in Figure 1.5 is zero because the plates are perfect conductors, which by definition cannot afford any voltage drop across their surfaces. Contrary to this example, electric fields may also change with time. For example, an electric field may be developed as a result of the motion of charged particles in the field.

Because the electric field can exert forces on charged particles, it is usually thought of as having a potential electric energy. The electric (or electrostatic) potential energy (in joules) is the potential energy associated with bringing together the building charges of a defined field from a state in which individual charges are completely separated. The potential energy of a field is directly proportional to the field amplitude. The line integral of an electric field intensity between two points, say A and B, is defined as the voltage between these two points, expressing the work done by the electric field to move an unit charge from point A to B. The following sections describe in detail how the electric energy can be calculated and how this energy contributes to microwave heating.

1.3.1.2 Magnetic field

A magnetic field is basically a field of forces generated by the movement of electric charges, a variation in the electric field in time, or the intrinsic magnetic field associated with the spin of particles. The concept of a magnetic field was first introduced by Hans Christian Ørsted (1820), who discovered that a conductor carrying a current produces a magnetic force field in its vicinity (Figure 1.6). His work was later followed by that of Andre Marie Ampere, who quantified the relationships involved.

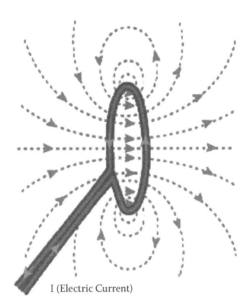

I (Electric Current)

Figure 1.6 Magnetic field lines in a circular loop.

Similar to the electric field, a magnetic field is a vector field with both a direction and a magnitude (strength). Magnetic fields may be in the form of a so-called B field or H field. The H field (also known as the magnetic field intensity and magnetic field strength) is normally considered a modification of the B field (also known as magnetic field induction and magnetic flux density) because of the magnetic fields produced by the material. Although, outside a material these two fields are indistinguishable, they may differ in magnitude and direction when present inside a material. These two fields are directly related to the magnetic property of the material known as permeability (defined in Section 1.4). In SI units, B fields and H fields are measured in teslas [(newton · second)/(coulomb · metre)] and ampere-turn per meter, respectively.

Similar to other fields, a magnetic field is commonly defined by its effect on the environment. A magnetic field is commonly defined in terms of the force it exerts on moving electric charges. A magnetic B field exerts a sideway force on a moving charged particle that is proportional to the strength of the field, particle charge, and the component of the velocity that is perpendicular to the magnetic field. The magnetic force is calculated as

$$Fm = qv \times B \tag{1.3}$$

where Fm is the magnetic force, B is the magnetic field in teslas, q is the electric charge of the particle, and v is the particle velocity. As can be deduced from Equation 1.3, the magnetic force should be orthogonal to both the v and B vectors, following the popular right-hand rule. Therefore, there would be no acceleration of the charged particles along the direction of motion in a magnetic field. This means that a magnetic field may only change the direction of the charge and not the associated kinetic energy. By combining Equations 1.1 and 1.3, the total force, known as the Lorentz force, exerted on a moving particle in an EM field may be obtained:

$$F = q(v \times B + E) \tag{1.4}$$

Magnetic fields have been knowingly or unknowingly used for many applications in ancient and modern societies. For instance, the fact that Earth generates its own magnetic field has been unconsciously used in navigation for a long time. The basic function of the compasses is to point toward the north pole of Earth's magnetic field, which is located near Earth's geographical north. On the other hand, an example of the application of magnetic fields in the modern world is the use of rotating magnetic fields in electric motors and generators.

1.3.2 Maxwell's equations

Electric and magnetic fields are not totally separate phenomena and are considered as two interrelated aspects of a single phenomenon called the EM field. In special relativity, the aspect of the EM field that is seen depends on the reference frame of the observer. This means that what an observer sees as an electric field may to another observer in a different frame of reference be considered as a mixture of electric and magnetic fields. The relationships between the fundamental EM quantities, including the field parameters and material dielectric properties, are explained by Maxwell's equations. In 1873, British physicist James Clerk Maxwell published his famous *Treatise on Electricity and Magnetism*, including the Maxwell equations, which embody and describe mathematically all phenomena of electromagnetism. Maxwell's equations are a set of equations, written in differential or integral form, stating the relationships between the fundamental EM quantities. These quantities are the electric field intensity E, the electric displacement or electric flux density D, the magnetic field intensity H, the magnetic flux density B, the current density J, and the electric charge density ρq. The vectors E and B are the basic field vectors defining the force exerted on a charge moving in an EM field as previously illustrated by Equation 1.4. The electric flux density vector D, also known as the displacement flux density, and the magnetic field intensity H take into account the dielectric and magnetic properties of the material media. The electric field in particular is a recurring parameter in microwave heating and is the prime source of energy transfer to the workload, which is discussed further in this book.

The differential form of Maxwell's equations is presented here because it leads to differential equations that can be used in the finite element analysis to study numerically the interaction between materials and EM fields. In Chapters 2–4, these equations are numerically solved to investigate the microwave interaction with concrete, mortar, and aggregates. For general time-varying fields, Maxwell's equations can be written as

$$\nabla \times H = J + \frac{\partial D}{\partial t} \qquad (1.5)$$

$$\nabla \cdot E = -\frac{\partial B}{\partial t} \qquad (1.6)$$

$$\nabla \cdot D = \rho_q \qquad (1.7)$$

$$\nabla \cdot B = 0 \qquad (1.8)$$

The first two equations are also referred to as Maxwell–Ampere's law and Faraday's law, respectively. Equations 1.7 and 1.8 are two forms of Gauss's law, the electric and magnetic forms, respectively. Another fundamental equation is the equation of continuity, which can be written as

$$\nabla \cdot J = \frac{\partial \rho_q}{\partial t} \qquad (1.9)$$

Of Equations 1.5–1.9, only three are independent. To obtain a closed system that can be solved to calculate the EM parameters for a specific field, the constitutive relations describing the macroscopic EM properties of the medium are to be included. They are given as

$$D = \varepsilon E = \varepsilon_r \varepsilon_0 E \qquad (1.10)$$

$$B = \mu H = \mu_r \mu_0 H \qquad (1.11)$$

$$J = \sigma E \qquad (1.12)$$

Here, ε is the permittivity of the medium, ε_r is the permittivity of medium relative to a vacuum, ε_0 is the permittivity of a vacuum, μ is the permeability of the medium, μ_r is the permeability of the medium relative to a vacuum, μ_0 is the permeability of a vacuum, and σ is the electrical conductivity of the medium. These properties are defined in the following section. By solving this set of equations subjected to the appropriate boundary conditions, the electric and magnetic parameters associated with any EM field can be calculated.

1.4 ELECTROMAGNETIC PROPERTIES

In Section 1.3.2, we briefly discussed the relationship between the EM properties of the materials and the field parameters. We found that, to obtain the field parameters for a specific problem, the EM properties of the constituent materials should be known. Here, we review the most fundamental EM properties of materials. Knowledge of the EM properties of materials in the microwave frequency range is essential for the proper design of microwave applicators.

Every material has a unique set of EM (dielectric) properties affecting the way in which the material interacts with the electric and the magnetic waves. Materials basically possess charge particles that give rise to three basic phenomena—conduction, polarisation, and magnetisation—when exposed to external fields. Depending on whether conduction, polarisation, or magnetisation is the most prominent phenomenon, the material may be classified as a conductor, dielectric, or a magnetic material, respectively. Microwave heating, especially microwave heating of concrete, deals normally with the dielectric materials in which polarisation caused by the external field is the prominent response. Therefore, the focus of this section is the properties of dielectric materials, especially dielectric materials acting as loads. Electric polarisation refers simply to the creation and alignment of dipoles within a material as a result of the relative shift of positive and negative electric charges in opposite directions within a dielectric under the influence of an electric field, making one side of the atom somewhat positive and the opposite side somewhat negative (Figure 1.7).

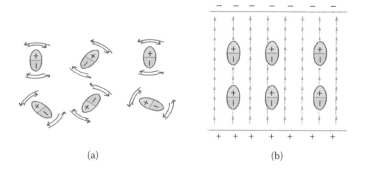

(a)   (b)

Figure 1.7 Dipolar molecules (a) in the absence and (b) in the presence of an electric field.

A dielectric material can be characterised by two independent EM properties, the complex permittivity ε and the complex permeability μ, which determine the response of the material to the electric and magnetic fields, respectively. Permittivity and permeability of materials can be determined through simple tests, which are briefly explained in the following sections. Once these fundamental properties have been obtained, the rest of the EM properties can be calculated using simple analytical relationships.

1.4.1 Permittivity and polarisation

The property that describes the behaviour of a dielectric under the influence of a high-frequency field is the complex permittivity. Basically, permittivity shows the material's ability to polarise in response to the field. Complex permittivity is defined as

$$\varepsilon = \varepsilon' - i\varepsilon'' \tag{1.13}$$

Here, ε' is the real part of complex permittivity, and ε'' is the imaginary part of complex permittivity. Dividing this by the permittivity of the free space ε_0, the property becomes dimensionless:

$$\frac{\varepsilon}{\varepsilon_0} = \frac{\varepsilon'}{\varepsilon_0} - i\frac{\varepsilon''}{\varepsilon_0} \tag{1.14}$$

Or,

$$\varepsilon_r = \varepsilon'_r - i\varepsilon''_r \tag{1.15}$$

where ε_0 (= 8.85×10^{-12} (F/m)) is the permittivity of the free space, ε_r is the relative permittivity, ε'_r is the dielectric constant, and ε''_r is the loss factor of the material. The dielectric constant is a measure of how much energy from an external electric field is stored in a material, and the loss factor is a measure of how dissipative or lossy a material is to an external field [17]. The ratio of the energy lost to the energy stored in a material is given as the loss tangent:

$$tan\delta = \frac{\varepsilon''}{\varepsilon'} = \frac{\varepsilon''_r}{\varepsilon'_r} \tag{1.16}$$

The electric susceptibility χ_e is a measure of how easily a dielectric material polarises when placed in an electric field and is directly related to the relative permittivity (ε_r) of the dielectric material.

$$\chi_e = \varepsilon_r - 1 \tag{1.17}$$

Electric susceptibility is defined as a proportionality constant relating the electric field's strength (E) to the induced polarisation density (P).

$$P = \varepsilon_0 \, E \, \chi_e \qquad\qquad (1.18)$$

As shown in Equation 1.17, the susceptibility and thus polarisability of a dielectric material such as concrete (the load) can be directly determined from relative permittivity. In addition, most of the other dielectric properties of materials such as reflectivity and attenuation, as will be discussed, are directly or indirectly related to the dielectric constant ε_r' and dielectric loss ε_r''. Therefore, experimental measurement of these two properties may provide much information about the behaviour of materials in EM fields.

Because of the complexities previously associated with the measurement of EM properties, there is little information available about the EM properties of concrete and other construction materials. However, the measurement of dielectric properties has been made straightforward recently, courtesy of many technical improvements. Currently, many commercial testing devices are available for the measurement of dielectric properties. In most of these devices, the measurement is as simple as placing a probe on the material surface and taking the measurements. Such devices use a number of simple techniques, including the open-ended coaxial probe method, resistivity cell method, parallel plate method, transmission line method, resonant cavity method, and free-space method [18]. Among these methods, the open-ended coaxial probe method is known to be the most suitable method for concrete because this method allows for the measurement of EM properties over a wide frequency range and is convenient for use with different sized concrete specimens. The principle of the probe method is that a measurement of the reflection from the material under test along with knowledge of its physical dimensions provide the information to characterise the permittivity and permeability of the material. A vector network analyser makes swept high-frequency stimulus-response measurements (Figure 1.8).

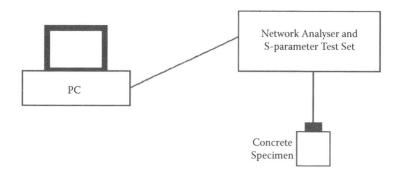

Figure 1.8 Open-ended coaxial apparatus for measurement of dielectric properties.

Dielectric properties of concrete and other construction materials are a function of a number of factors, such as the constituent materials, mix proportions, water content, microwave frequency, and temperature. Therefore, it is important to measure the properties of such materials under conditions similar to those expected in practice. Among these factors, water content and microwave frequency are the two most important parameters influencing dielectric properties. The variation in concrete's dielectric constant and loss factor for a typical concrete specimen with water content and microwave frequency are shown in Figures 1.9 and 1.10. As seen, both the

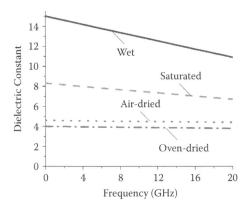

Figure 1.9 Dielectric constant of a typical concrete at microwave frequency range. (From Rhim, H.C. and Buyukozturk, O. Electromagnetic properties of concrete at microwave frequency range. *ACI Materials J*, 1998, **95**:262–271. With permission.)

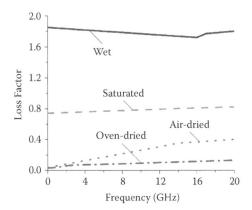

Figure 1.10 Loss factor of concrete at microwave frequency range. (From Rhim, H.C. and Buyukozturk, O. Electromagnetic properties of concrete at microwave frequency range. *ACI Materials J*, 1998, **95**:262–271. With permission.)

dielectric constant and loss factors of the concrete increase significantly with an increase in microwave frequency or the water content of concrete, leading to a higher heating potential when exposed to microwaves.

1.4.2 Electric conduction

All materials, including dielectrics, possess conductivity to some extent. Moreover, conductors are commonly used in the fabrication of microwave heating chambers and transmission lines [19]. Therefore, understanding the conduction phenomenon is necessary for understanding the microwave heating process. When a conductor material is placed in an electric field, because of the movement of free electrons inside the material under the influence of the electric field, a current known as the conduction current is produced in the material. In linear isotropic conductors, the conduction current density (J, A/m^2) is related to the electric field intensity through Equation 1.12, repeated here.

$$J = \sigma E$$

where σ is the conductivity and is measured in siemens per meter units (S/m). The electric conductivity is related to permittivity through the following equation:

$$\sigma = \varepsilon''\omega = (\varepsilon' \tan\delta)\omega = (\varepsilon'_r\varepsilon_0 tan\delta)(2\pi f) \tag{1.19}$$

where f and ω are the frequency and angular frequency of the EM wave, respectively. The electric conductivity of concretes with various water contents at microwave frequency ranges, calculated from the loss factor using Equation 1.19, is shown in Figure 1.11. As can be seen, because of the high

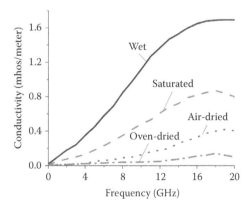

Figure 1.11 Conductivity of concrete calculated using Equation 1.19.

conductivity of water, the conductivity of concrete increases significantly with its water content.

1.4.3 Reflectivity and transmissivity

The mismatch between the dielectric constants at the interface between two different media causes some of the incident waves to be reflected and the rest to be transmitted into the next medium (Figure 1.12). The reflection coefficient R can be calculated from

$$R = \frac{\sqrt{\varepsilon_{r2}} \, \cos\theta_i - \sqrt{\varepsilon_{r1}} \, \cos\theta_t}{\sqrt{\varepsilon_{r2}} \, \cos\theta_i + \sqrt{\varepsilon_{r1}} \, \cos\theta_t} \qquad (1.20)$$

where ε_{r1} is the dielectric constant for medium 1, ε_{r2} is the dielectric constant for medium 2, θ_i is the angle of incidence, and θ_t is the angle of transmission. The square of reflection ($|R|^2$), called *reflectivity* and denoted as r, is the ratio between the portion of the incident EM power that is reflected to the first medium and the total incident EM power. Therefore, the transmissivity c, which is the ratio between the power transmitted to the second medium and the total EM power, may be obtained from

$$c = 1 - r \qquad (1.21)$$

The transmissivity coefficient of concretes with various water contents at microwave frequency ranges, calculated from the loss factor obtained by using Equations 1.20 and 1.21, is shown in Figure 1.13. As can be seen, the transmissivity coefficient of concrete decreases significantly with an increase in the water content of concrete. This means that a higher proportion of the incident power may be transmitted into a concrete component

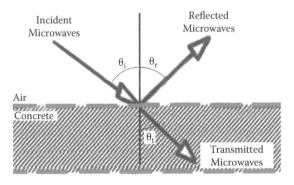

Figure 1.12 Reflection and transmission of the waves at the air-concrete interface.

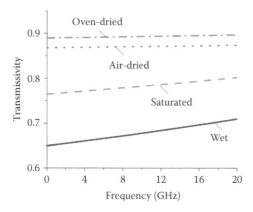

Figure 1.13 Transmissivity of concrete.

with lower water content than similar concrete components with higher water contents.

1.5 MICROWAVE HEATING MECHANISM

In the previous sections, we explained the basic constituents of the EM waves and the way they interact with one another. In this section, you will learn how an EM field may be used to heat the materials.

The microwave heating phenomenon may be generally thought of as occurring in two stages. The first is the EM aspect, concerned with the propagation of radiation and power absorption by the medium. Second, the energy absorbed is converted into heat, which is transported by convection, conduction, and radiation.

In general, the EM energy propagating in a dielectric material may be dissipated by two primary mechanisms. In the first mechanism, known as direct current (DC) conductivity loss, the dielectric material behaves like a poor electrical conductor, having a finite resistivity measurable at DC. This type of loss is usually substantially constant as the frequency extends upward.

The other mechanism is called dielectric loss or polarisation caused by an altering electric field. When a dielectric material such as concrete is exposed to an electric field, electric charges do not flow through the material but only move slightly away from their average equilibrium positions. In such a state, known as dielectric polarisation, the positive charges move along the field while the negative charges move in the opposite direction. Dipolar components of the molecules in the dielectric material couple electrostatically to the electric field and tend to align themselves with the field mechanically (Figure 1.7). Because the electric field is altering with time, the dipoles will attempt to realign each time the field reverses and so are in a constant state of mechanical oscillation in tandem with the microwave frequency. Frictional forces generated within the molecules cause heat to be developed because of the motion of the dipoles.

The amount of microwave energy dissipated through dielectric loss varies significantly with the polarisability and other dielectric properties of the material heated. Water is the most popular material displaying such polar characteristics and is the main component in most of the dielectrics that show good microwave power absorption. Concrete, cementitious mortar, and some aggregates are also porous materials with pores that may be filled with water. Water absorbs EM energy very strongly when highly pure, although it also has a high DC electrical resistivity of approximately 105 Ω. However, when small quantities of solid are dissolved in water, the DC resistivity falls, and conduction via movement of the ionic charge carriers may become a significant component in heat dissipation.

It is interesting to note that, unlike dipolar heating, conduction heating tends to fall away when raising the frequency into the microwave domain. This is mainly because the mass of the ions is such that their movement is curtailed. Therefore, polarisation caused by the alternating field may be considered the main heating mechanism involved in the microwave heating of dielectric materials such as concrete and its constituent materials.

1.6 ELECTROMAGNETIC POWER TRANSFER

We discussed previously that exposure to microwave power may lead to heating of the dielectric materials through dipolar losses. However, it is important to understand how the power generated in the microwave generator unit is transferred to the material. In microwave heating, a hollow metallic tube of either rectangular or circular cross section, made of aluminum, copper, or brass of various sizes, is usually used to transmit the generated power. Such a structure is commonly known as a waveguide (Figure 1.14). Waveguides may be used to transfer the microwave power directly to the material surface or to a microwave heating chamber (cavity) in which the material is placed. Microwave cavities are normally in the form of a metallic chamber similar to those used in domestic microwave ovens. There are sets of standard waveguide dimensions for the range of microwave frequencies and waveguide cross sections used. For instance, the standard dimensions of rectangular waveguides at the common frequencies used in microwave heating are listed in Table 1.1.

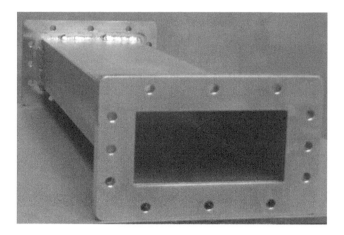

Figure 1.14 A rectangular waveguide.

Table 1.1 Standard rectangular waveguide dimensions at the frequencies commonly used in microwave heating

Frequency (GHz)	Designation	Width, mm (in.)	Height, mm (in.)
0.986	WR770	195.58 (7.70)	85.85 (3.38)
2.45	WR340	86.36 (3.4)	43.18 (1.7)
10.6	WR90	22.86 (0.9)	10.16 (0.4)
18	WR42	10.66 (0.42)	4.31 (0.17)

An infinite number of distinct EM configurations, which determine the distribution of power within the waveguide or cavity and on the incident surface of the material, may exist inside each type of waveguide/cavity, depending on the cross-sectional dimensions, frequency, and properties of the dielectric materials placed inside the waveguide/cavity. These configurations are normally known as "modes of propagation" and correspond to the solutions of Maxwell's equations. These configurations may be categorised under several general types. Among the various types, transverse electric (TE) mode and transverse magnetic (TM) mode are the most commonly excited modes in microwave heating systems.

In TE modes, the electric field vector is always perpendicular to the waveguide axis. At this mode, the electric field vector in the propagation direction is zero. Contrary to the TE mode, in TM modes, the magnetic field vector is always perpendicular to the waveguide axis. In this mode, the magnetic field vector in the propagation direction is zero.

It is possible to have several modes of propagation inside a particular waveguide or cavity. The mode that has the lowest frequency for a particular waveguide is known as the dominant mode. Cavities with more than one dominant mode are known as multimode cavities. The waveguides used for microwave heating generally have dimensions such that only the dominant mode propagates along it.

The most fundamental modes in the TE and TM categories that are usually excited by the microwave heating systems are the TE_{10} and TM_{11} modes. The TE_{10} mode is the usual choice for single-mode commercial waveguides. It does not vary in one of the transverse directions and has a sinusoidal distribution in the other. Hence, it represents a heating problem similar to that of heating a two-dimensional slab. The form of the incident TE_{10} mode may be represented as

$$E = (0,0,E_z) = \left\{0,0,\sin\left[\pi(a - y)/2a\right]\right\} \qquad (1.22)$$

where $2a$ is the width of the waveguide. The concepts of wave propagation in waveguides as well as the multimode cavities used for the heating of construction materials are presented in Chapter 6.

1.7 PENETRATION DEPTH AND ATTENUATION FACTOR

As a result of the energy losses, the microwave energy attenuates as it propagates in the dielectric material. Figure 1.15 shows the microwave electric field distribution inside a 10-mm thick saturated concrete block when subjected to 1 W of microwave power through direct contact with a standard WR340 rectangular waveguide (Figure 1.16). It can be seen that, as a result

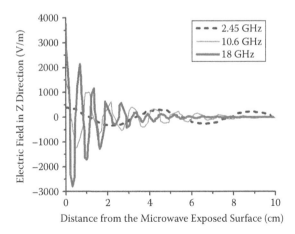

Figure 1.15 Distribution of electric field in a saturated concrete at various microwave frequencies.

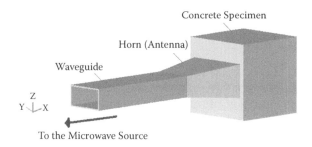

Figure 1.16 A concrete specimen subjected to microwaves through a rectangular waveguide.

of power attenuation, the penetration of microwave power in the dielectric material is usually limited to a specific depth, depending on the microwave frequency and dielectric properties of the material.

The power penetration depth is defined as the depth at which the transmitted power drops to $1/e$ of its value at the surface and is given by

$$d_p = \frac{1}{2\beta} \tag{1.23}$$

where β is the attenuation factor. Having measured the basic EM properties of a dielectric material, its attenuation factor at a specific frequency may be calculated as

$$\beta = \frac{2\pi f}{c} \sqrt{\frac{\varepsilon_r' \left(\sqrt{1 + \tan^2 \delta} - 1 \right)}{2}} \tag{1.24}$$

where c is the speed of light, $\tan \delta$ is the loss tangent, ε_r' is the dielectric constant, and f is the microwave frequency. The attenuation factor of concrete (calculated using Equation 1.24) at various water contents and microwave frequencies is shown in Figure 1.17. As can be seen, because of the increase in the power losses, the attenuation factor of concrete increases significantly with an increase in the microwave frequency or water content of concrete.

1.8 FORMULATION OF MICROWAVE POWER DISSIPATION (DIELECTRIC LOSS)

We discussed previously that in the microwave frequency range, the dielectric losses become the main contributing factor in dielectric heating. In

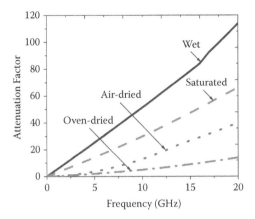

Figure 1.17 Attenuation factor of concrete.

the following, we first formulate the total energy potential of a microwave field. Then, we review a number of the methods available for estimating the portion of this energy that is absorbed and dissipated in a material placed in a microwave field. The electric W_e and magnetic W_m energies may be generally defined as

$$W_e = \int_V \left(\int_0^D E \cdot dD \right) dV = \int_V \left(\int_0^T E \cdot \frac{\partial D}{\partial t} dt \right) dV \tag{1.25}$$

$$W_m = \int_V \left(\int_0^B H \cdot dB \right) dV = \int_V \left(\int_0^T H \cdot \frac{\partial B}{\partial t} dt \right) dV \tag{1.26}$$

The time derivatives of these expressions are the electric and magnetic power, respectively:

$$P_e = \int_V E \cdot \frac{\partial D}{\partial t} dV \tag{1.27}$$

$$P_m = \int_V H \cdot \frac{\partial B}{\partial t} dV \tag{1.28}$$

These quantities are related to the resistive and radiative energy, or energy loss, through Poynting's theorem:

$$-\int_V \left(E\frac{\partial D}{\partial t} + H\frac{\partial B}{\partial t} \right) dV = \int_V J \cdot E\, dV + \oint_s (E \times H) \cdot n\, ds \qquad (1.29)$$

where V is the computation domain, and s is the closed boundary of V. The quantity $E \times H$ is called the Poynting vector.

$$S = \frac{1}{2} E \times H \qquad (1.30)$$

By substituting Equations 1.10, 1.11, 1.12, and 1.19 into Equation 1.29, with some rearrangements, we may obtain

$$\oint_s S \cdot n\, ds = -\frac{1}{2}\omega\varepsilon_0\varepsilon_r'' \int_V E \cdot E\, dV + i\omega \int_V \left(\frac{\mu_0}{2} H \cdot H + \frac{\varepsilon_0\varepsilon_r'}{2} E \cdot E \right) dV \qquad (1.31)$$

which states that the net power flow across a surface s enclosing a volume V equates to the power dissipated in the medium (real part) and that stored in the electric and magnetic fields (imaginary part). By applying the divergence theorem to Equation 1.31, the point form of Poynting theorem will be obtained as follows:

$$\nabla \cdot S = -\frac{1}{2}\omega\varepsilon_0\varepsilon_r''E \cdot E + i\omega\left(\frac{\mu_0}{2} H \cdot H + \frac{\varepsilon_0\varepsilon_r'}{2} E \cdot E \right) \qquad (1.32)$$

Hence, the power dissipated per unit volume (real part) can be written as

$$P^M(r) = -Re(\nabla \cdot S) = \frac{1}{2}\omega\varepsilon_0\varepsilon_r''|E|^2 \qquad (1.33)$$

Equation 1.33 shows that the radiative microwave power dissipated per volume is directly proportional to the square of the electric field intensity's norm. Hence, if one could use Maxwell's equations to estimate the electric field's intensity in a material, the power dissipated in that material may be easily estimated using Equation 1.33. However, owing to the complexity of Maxwell's equations, approximations are normally used to estimate microwave power dissipation in dielectric materials. In the following section, one of the most popular power dissipation estimation techniques is presented. The variation of the norm of electric field with distance from the incident surface of a saturated concrete subjected directly to 1 W of microwave power through a rectangular waveguide (Figure 1.16) at various

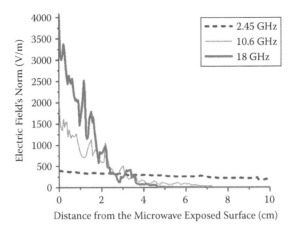

Figure 1.18 Variation of the electric field norm within a saturated concrete specimen.

frequencies is shown in Figure 1.18. As can be seen, the electric field norm and thus the microwave power in the concrete seem to attenuate exponentially within a concrete specimen. In addition, the penetration depth decreases substantially with an increase in microwave frequency.

1.8.1 Lambert's law

Lambert's law is the most common approximation used to estimate microwave power dissipation. This method considers an exponential decay of microwave energy absorption inside the dielectric material. The original form of Lambert's law may be written as

$$I(x) = I_0 e^{-2\beta x} \tag{1.34}$$

where $I(x)$ is the transmitted power flux into the medium, and I_0 is the incident power of microwaves. Equation 1.34 may be used to predict the variations in the microwave energy with distance x from the microwave-exposed surface of the material. Hence, the microwave energy dissipated at a specific point inside a dielectric material may be estimated by differentiating $I(x)$ with respect to x which results in

$$PL(x) = -\frac{\partial I(x)}{\partial x} = 2\beta I_0 e^{-2\beta x} \tag{1.35}$$

Lambert's law, when applicable, is easy to use. However, prior to using Lambert's law for microwave power estimation in a specific microwave heating application, its validity under the specified conditions should be verified. It has been debated that, although Lambert's law is valid for samples or loads thick enough to be treated as infinitely thick, it offers a poor approximation when applied in many practical situations. For instance, in applications where the object's dimensions are typically of the order of the wavelength of the radiation, the power distribution is very different from those predicted by Lambert's law. In such problems, usually, standing waves in the sample give rise to local hot spots where the power achieves local maxima. Also, corners and edges are known to focus radiation.

The validity of Lambert's law for estimating the microwave energy distribution in a few materials has been investigated in available literature. For instance, Ayappa et al. compared numerical model predictions using Maxwell and Lambert's law to represent power dissipation in slabs [1]. They obtained a critical thickness above which the use of Lambert's approximation is valid and showed that the two formulations predict identical power profiles for slabs at least 2.7 times thicker than the penetration depth. By taking this into consideration, the minimum thicknesses of concrete, at various microwave frequencies and concrete water contents, required to ensure the validity of Lambert's law for the estimation of microwave power in concrete are listed in Table 1.2. As can be seen, while for the saturated and wet concretes the minimum thickness required is less than the thickness of typical concrete components used for structural and nonstructural applications in typical buildings, this is not always the case when the water content of concrete decreases.

There are two other important points that must be noted when using Lambert's law. First, the incident microwave power considered in Lambert's law does not take into account the reflections at the interface between the material and the free space. Therefore, the use of Lambert's law requires an estimate of the transmitted power density [20]. Second, Lambert's law considers a uniform distribution of microwave power on the incident surface of the material and does not account for the power distribution associated with the dominant microwave mode. To use Lambert's law for microwave power estimation in actual applications, the original formulation should be modified to account for these two factors.

Table 1.2 Minimum thickness of concrete components to guarantee the validity of Lambert's law

Frequency (GHz)	Moisture condition: Minimum thickness of concrete block, cm (in.)			
	Wet	Saturated	Air-dried	Oven-dried
0.96	26.95 (10.61)	49.67 (19.55)	464.88 (183.02)	644.94 (253.91)
2.45	10.96 (4.31)	19.97 (7.86)	122.53 (48.24)	161.30 (63.50)
10.6	2.47 (0.97)	4.23 (1.66)	9.40 (3.70)	25.09 (9.88)
18	1.36 (0.53)	2.31 (0.91)	3.93 (1.55)	11.40 (4.49)

One way to account for the reflections at the material-ambient interface is to multiply the incident microwave power by the transmissivity coefficient as defined by Equation 1.21:

$$PL(x) = -\frac{\partial I(x)}{\partial x} \times c = 2\beta I_0 e^{-2\beta x} \times c \tag{1.36}$$

In addition, to account for the distribution of microwave power at the material interface, the incident power should be multiplied by an appropriate function that simulates the power distribution associated with the dominant microwave mode. For instance, by considering the TE_{10} mode as the dominant microwave mode in typical rectangular waveguides used in industrial microwave heating, Lambert's law should be modified to account for the sinusoidal power distribution associated with this mode.

We previously showed that the radiative microwave power dissipated per unit volume of the dielectric material is directly proportional to the square of the electric field intensity's norm (Equation 1.33). Keeping this in mind, we may conclude that the power dissipated in a dielectric material subjected directly to microwaves with a TE_{10} dominant mode should follow a sine² distribution. Therefore, prior to using Lambert's law for estimating the variation in microwave power at the TE_{10} mode, the original form of Lambert's law may be modified as

$$I(x) = Psin^2\left(\pi\frac{a-y}{2a}\right) \times e^{-2\beta x} \tag{1.37}$$

where $2 \times a$ is the waveguide width, and P is the peak of the sine² function. The peak value of the sine² distribution may be obtained by equating the area under the sine² distribution to the area under a uniform (rectangular) power distribution with the intensity of I_0 as assumed in the original form of Lambert's law:

$$2aI_0 = 2 \times \int_0^a Psin^2\left(\pi\frac{a-y}{2a}\right)dy \Rightarrow P = 2I_0 \tag{1.38}$$

Hence, the final power dissipation function accounting for the power distribution associated with the TE_{10} mode as well as the reflection at the material interface with ambient may be considered as

$$PL(x) = -\frac{\partial I(x)}{\partial x} \times c = 2\beta \times 2I_0 \times c \times sin^2\left(\pi\frac{a-y}{2a}\right) \times e^{-2\beta x} \tag{1.39}$$

Comparing Equation 1.39 with Equation 1.35 illustrates that the incident microwave power delivered through a TE_{10} mode waveguide has a peak incident power that is twice the equivalent uniform incident power. Thus, the maximum peak temperature expected locally as a result of heating of the concrete with a TE_{10} microwave mode is expected to be twice the peak temperature reached when uniform microwave heating is assumed. This simply illustrates the importance of accounting for the incident microwave power distribution in power estimations.

1.9 HEAT TRANSFER AND TEMPERATURE RISE IN MICROWAVE HEATING OF CONCRETE

As explained in Section 1.5, the microwave power absorbed by the material or load leads to heating and temperature rise. To understand the phenomena involved in the microwave heating of concrete, as discussed in the following chapters, it is important to predict the temperature rise in concrete components exposed to microwaves.

It was shown previously that microwave power dissipation in concrete can be calculated either using the Poynting theorem when the distribution of the electric field is obtained by solving Maxwell's equations or using the approximate equations, such as Lambert's law. Once the value of the microwave power dissipated in the material is known, the heat transfer equation may be solved to predict the temperature rise. The basic heat transfer equation for concrete exposed to microwaves may be written as

$$\frac{\partial}{\partial t}(\rho C)T + \nabla \cdot q = P(x) \tag{1.40}$$

where T is the temperature; ρ is the mass density of concrete; C is the specific heat of concrete, is the gradient operator; q is the total heat flux vector, including the conductive heat flux q_{cd} and the convective heat flux q_{cv}; and $P(x)$ is the distributed source of heat, which can be estimated using either the Poynting theorem or Lambert's law. The conductive heat flux q_{cd} may be expressed as linear combinations of the gradient of temperature T:

$$q_{cd} = -k\nabla T \tag{1.41}$$

where k is the heat conductivity. Moreover, heat may also be transferred through the movement of water in the concrete:

$$q_{cv} = C_w TJ \tag{1.42}$$

where C_w is the specific heat of water, and J is the water flux. The water flux J can be expressed in terms of the gradient of pore pressure:

$$J = -\frac{a}{g}\nabla P \tag{1.43}$$

where a is the water permeability of concrete, and g is Earth's gravity. Hence,

$$q_{cv} = C_w TJ = -a\left(\frac{C_w}{g} T\nabla P\right) \tag{1.44}$$

However, because the water permeability of concrete is about three orders of magnitude smaller than heat conductivity, this term can be neglected; thus, Equation 1.39 may be simplified as

$$\rho C \frac{\partial T}{\partial t} = -\nabla \cdot \left(-k\nabla T\right) + P(x) \tag{1.45}$$

It is noteworthy that Equation 1.43 is the classical Darcy's law; hence, its validity is known to be limited to saturated porous materials. However, previous studies have shown that Darcy's law is applicable to a heated unsaturated material provided that P is interpreted as the pressure of water vapour in the pores rather than the pressure of liquid capillary water [21].

Once the heat source $P(x)$ is known, Equation 1.45 may be used to calculate the temperature distribution inside the microwave-heated concrete.

1.10 MASS TRANSFER PHENOMENON AND THE PORE PRESSURE DEVELOPMENT IN MICROWAVE-HEATED CONCRETE

Another phenomenon that has to be fully comprehended for a better understanding of the applications of microwave heating in concrete technology is the way microwave heating affects the movement of water and water vapour in the concrete pores. As a result of microwave heating of concrete, part of the pore water may turn into vapour. If the vapour generation rate exceeds the rate of the vapour migration from the surface of concrete, substantial pore water pressure may be developed within the concrete, which may affect its structural integrity. Understanding this phenomenon and predicting the behaviour of water and vapour in concrete requires a detailed understanding of the mass transfer phenomenon in concrete. The mass conservation equation for concrete may be expressed as

$$\frac{\partial w}{\partial t} + \nabla \cdot J = HD(w) \tag{1.46}$$

Here, specific water content, time, gradient operator, water flux vector, and change in free-water content because of hydration and dehydration. The free-water content in the liquid phase within the concrete can be determined by means of the so-called Equation of state of pore pressure [22]. For temperatures above the critical point of water (374.15°C), all free water is assumed to have been vapourised; thus, there is no liquid phase. For temperatures below the critical point of water, the free-water content depends on the temperature and the ratio of water vapour pressure to saturation vapour pressure [23]. In the following, the semiempirical expressions from Reference 22 are used. For nonsaturated concrete, the following formula has been proposed:

$$\frac{w}{C} = \left(\frac{W_{s1}}{C} \times h \right)^{\frac{1}{m(T)}} \quad h \leq 0.96 \tag{1.47}$$

Here, is the water content of concrete, W_{s1} is the saturation water content at 25°C, C is the mass of (anhydrous) cement per cubic metre of concrete,

$$h = \frac{P}{P_s(T)}$$

where $P_s(T)$ = saturation pore pressure at temperature T, and $m(T)$ is an experimentally determined empirical expression as follows [22]:

$$m(T) = 1.04 - \frac{(T+10)^2}{22.3(25+10)^2 + (T+10)^2} \tag{1.48}$$

For saturated concrete, the ratio of free water to cement is determined by [22]

$$\frac{w}{C} = \frac{W_{s1}}{C} \left[1 + 0.12(h - 1.04) \right] \quad h \geq 1.04 \tag{1.49}$$

The transition between $h = 0.96$ and $h = 1.04$ is assumed to be linear between the values of $w_{0.96}$ and $w_{1.04}$; thus,

$$w = w_{h=0.96} + (h-0.96) \times \frac{w_{h=1.04} - w_{h=0.96}}{1.04 - 0.96} \quad 0.96 \le h \le 1.04 \tag{1.50}$$

The saturation pore pressure at different temperatures can be calculated using the following semiempirical equation [24]:

$$Ln\left(\frac{P_s}{P_c}\right) = \frac{T_c}{T}\left[a(1-T_r) + b(1-T_r)^{1.5} + c(1-T_r)^3 + d(1-T_r)^6\right] \tag{1.51}$$

where $T_r = T/T_c$, $T = 647.7$ K, $P_c = 22.07$ MPa, $a = -7.7645$, $b = 1.45938$, $c = -2.7758$, and $d = -1.23303$. By substituting Equations 1.43, 1.47, 1.49, and 1.50 into Equation 1.46, the following set of equations may be obtained and can be used for estimating the pore water pressure development in a microwave-heated concrete component:

$$\frac{\partial}{\partial t}\left(C\left(\frac{W_{s1}}{C} \times \frac{P}{P_s(T)}\right)^{\frac{1}{m(T)}}\right) + \nabla \cdot \left(-\frac{a}{g}\nabla P\right) = HD(w), \tag{1.52}$$

$$\frac{P}{P_s(T)} \le 0.96$$

$$\frac{\partial}{\partial t}\left(W_{\frac{P}{P_s(T)}} = 0.96 + \left(\frac{P}{P_s(T)} - 0.96\right) \times \frac{W_{\frac{P}{P_s(T)}} = 1.04 - W_{\frac{P}{P_s(T)}} = 0.96}{1.04 - 0.96}\right)$$
$$+ \nabla \cdot \left(-\frac{a}{g}\nabla P\right) = HD(w), \ 0.96 \le \frac{P}{P_s(T)} \le 1.04 \tag{1.53}$$

$$\frac{\partial}{\partial t}\left(W_{S1} + \left(1 + 0.12\left(\frac{P}{P_s(T)} - 1.04\right)\right)\right) + \nabla \cdot \left(-\frac{a}{g}\nabla P\right) = HD(w), \tag{1.54}$$

$$\frac{P}{P_s(T)} \ge 1.04$$

At ambient temperatures, the permeability of concrete is controlled by nanopores, which explains the extremely low permeability at these temperatures. But, this is not the case at high temperatures. When the temperature is increased to above 100°C, the permeability rises sharply. This can be explained by heat-induced changes in the structure of small pores [25]. The following expressions have been proposed to predict the variation in permeability with temperature [25]:

for $T \leq 100°C$ (1.55)

and

for $T \geq 100°C$ (1.56)

where reference permeability at $T = 25°C$, $f_1(h)$, $f_2(T)$, and $f_3(T)$ are functions described by following expressions:

$$f_1(h) = \alpha + \frac{1-\alpha}{1+\left(\frac{1-h}{1+h_c}\right)^4} \quad \text{for } h \leq 1, \text{ and } f_1(h) = 1 \quad \text{for } h \geq 1 \quad (1.57)$$

where

$$\alpha = 1/\left[1+0.253\left(100 - min\left(T, 100\ °C\right)\right)\right]$$

and

$$h_c = 0.75.$$

$$f_2(T) = exp\left(\frac{Q}{R_g}\left(\frac{1}{T_0} - \frac{1}{T}\right)\right) \quad (1.58)$$

$$f_3(T) = 5.5\left[\frac{2}{a + exp\left(-0.455\left(T - 100\right)\right)} - 1\right] + 1 \quad (1.59)$$

where absolute temperature, Q = activation energy for water migration, and R_g = gas constant. Having taken this information into consideration, the pore pressure and the moisture distribution in the concrete specimen subjected to microwave heating can then be obtained by coupling the heat and mass transfer equations.

1.11 MICROWAVE HEATING SAFETY

So far, we have briefly discussed how microwave heating can be used to heat dielectric materials and how this may be used in civil engineering applications. However, before we proceed further into the details of these applications, it is essential to make sure that such techniques can be safely deployed at civil engineering work sites. Safety is an important aspect of any industrial process. All industrial machinery should be designed to ensure safety of operators. Previous studies conducted on the safety of microwave heating showed that microwave heating can be as safe as conventional heating. The two main safety concerns about microwave heating are briefly explained in the following sections.

1.11.1 Radiation hazards

It is well known that, unlike γ- and x-ray frequencies, the photonic energy at the various microwave frequencies is not sufficient to ionise the exposed material. Available literature does not show any confirmed relationship between exposure to EM waves at the various microwave frequencies and development of various cancers [26]. The only proven effect of microwave heating on biological materials is thermal heating. Hence, thermal injury may be considered the only confirmed health hazard associated with microwave heating. The danger of thermal injury caused by exposure to microwaves increases with an increase in microwave frequency and power. As illustrated in Chapters 2 to 5, higher frequencies lead to more concentrated heating because of their shorter penetration depth. Thus, a much lower power may be needed to cause burns when exposed to higher frequencies.

Shielded chambers are normally used to protect operators against microwave radiation. Chapter 6 discusses further on how the proper design of microwave heating chambers can ensure the safety of operators. Usually, leakage meters are deployed on a periodic basis to detect microwave leakage in microwave heating processes. On detection of excess emissions, the equipment should be immediately (usually automatically) shut down and the cause of emissions investigated. Under the International Electrotechnical Commission (IEC) standard, which is applicable to equipment operating in the frequency range from 300 MHz to 300 GHz, power density is measured at least a distance of 5 cm from any accessible location away from the equipment and should be limited to 5 mW/cm^2 during "normal" operation and 10 mW/cm^2 during "abnormal" operations to ensure the safety of users.

1.11.2 High-voltage hazards

Although most of the public concern about microwave heating has been focused on health hazards arising from exposure to EM radiation, experience shows that the hazards associated with the exposure to high-voltage microwave generators is, at least, of equal importance. In almost all microwave generators, depending on the design, very high voltages, up to tens of thousands volts, may be present during operation. High voltages may even be present during the nonoperational stage as most microwave generators include one or more capacitors to store high-voltage electric charges. These high voltages can cause serious injury or even death if the human body becomes a part of the electrical circuit because of inadvertent contact or when it is placed in close proximity to the high-voltage conductors or capacitors. To address these issues, all microwave-generating equipment should be enclosed properly and be accessible only to well-trained experts. Most available commercial microwave systems are designed for this through the use of door interlocks that shut the power down and discharge the stored charges automatically once the doors of the generators are opened.

1.12 SUMMARY

Because of its higher efficiency, less-polluting nature, and unique capabilities in adjusting the pattern of heating, microwave heating is rapidly replacing conventional heating technologies used in many industries, including food processing, timber drying, ceramic fabrication and processing, mining, and so on. Similarly, microwave heating has been recently proposed for a number of innovative applications in the concrete industry. Typically, fresh and hardened concrete are heated to different extents and degrees of uniformity when exposed to microwaves of the appropriate frequency range. Examples of actual applications in recent years include microwave curing of prefabricated concrete elements, microwave-assisted demolition of concrete structures, drilling and cutting of concrete, and microwave-assisted concrete recycling, which is discussed in detail in the following chapters. To provide the background knowledge required to understand the working principles of the microwave-assisted methods introduced in Chapters 2 to 4, this chapter started by explaining the fundamentals of microwave heating and the microwaves–concrete interaction phenomena. After reviewing some of the basic EM properties of concrete, a number of analytical methods for the estimation of microwave power dissipation and resulting temperature rise and pore pressure development within the concrete were reviewed. Furthermore, the general safety and efficiency concerns associated with microwave heating were discussed.

REFERENCES

1. Ayappa, K.G., Davis, H.T., et al., Microwave heating: An evaluation of power formulations. *Chemical Engineering Science*, 1991, **64**(4):1005–1016.
2. Metaxas, A.C. and Meredith, R.J., *Industrial Microwave Heating*. Stevenage, UK: Peter Perigrinus, 1983, 354.
3. Akbarnezhad, A. and Ong, K.C.G., Microwave decontamination of concrete. *Magazine of Concrete Research*, 2010, **62**(12):879–885.

4. Akbarnezhad, A., Ong, K.C.G., et al., Effects of the parent concrete properties and crushing procedure on the properties of coarse recycled concrete aggregates. *Journal of Materials in Civil Engineering*, 2013, **25**(12):1795–1802.
5. Leung, C.K.Y. and Pheeraphan, T., Determination of optimal process for microwave curing of concrete. *Cement and Concrete Research*, 1997, **27**(3):463–472.
6. Leung, C.K.Y. and Pheeraphan, T., Very high early strength of microwave cured concrete. *Cement and Concrete Research*, 1995, **25**(1):136–146.
7. Rattanadecho, P., Suwannapum, N., et al., Development of compressive strength of cement paste under accelerated curing by using a continuous microwave thermal processor. *Materials Science and Engineering*, 2008, **427**:299–307.
8. White, T.L., Grubb, R.G., Pugh L.P., Foster, D., Jr., and Box, W.D., Removal of contaminated concrete surface by microwave heating—phase 1 results. *Proceedings of 18th American Nuclear Society Symposium on Waste Management*, Tucson, AZ, March 1–5, 1992. Available at http://inis.iaea.org/search/search.aspx?orig_q=RN:23044093
9. Spalding, B., Volatility and extractability of strontium-85, cesium-134, cobalt-57, and uranium after heating hardened Portland cement paste. *Environmental Science Technology*, 2000, **34**:5051–5058.
10. Zi, G. and Bažant, Z.P., Decontamination of radionuclides from concrete by microwave heating. II: Computations. *Journal of Engineering Mechanics*, 2003, **129**(7):785–792.
11. Bažant, Z.P. and Zi, G., Decontamination of radionuclides from concrete by microwave heating. I: Theory. *Journal of Engineering Mechanics*, 2003, **129**(7):777–784.
12. Sri Ravindrarajah, R. and Tam, C.T., Properties of concrete made with crushed concrete as coarse aggregate. *Magazine of Concrete Research*, 1985, **37**(130):29–38.
13. Tam, V.M. and Tam, C.M., *Re-use of Construction and Demolition Waste in Housing Developments*. New York: Nova Science, 2008, viii.
14. Hasted, J.B. and Shah, M.A., Microwave absorption by water in building materials. *British Journal of Applied Physics*, 1964, **15**:825–836.
15. Chi-Sun, P. and Dixon, C., The use of recycled aggregate in concrete in Hong Kong. *Resources, Conservation and Recycling*, 2007, **50**(3):293–305.
16. De Juan, M.S. and Gutierrez, P.A., Study on the influence of attached mortar content on the properties of recycled concrete aggregate. *Construction and Building Materials*, 2009, **23**(2):872–877.
17. Rhim, H.C. and Buyukozturk, O., Electromagnetic properties of concrete at microwave frequency range. *ACI Materials Journal*, 1998, **95**:262–271.
18. Pozar, D.M., *Microwave Engineering*. New York: Wiley.
19. Akbarnezhad, A., Ong, K.C.G., Chandra, L.R., and Lin, Z.S., Optimized deconstruction of buildings using building information modeling. Paper presented at *Construction Research Congress, ASCE*, Purdue University, West Lafayette, IN, 2012.
20. Stuchly, S.S. and Hamid, M.A.K., Physical parameters in microwave heating processes. *Journal of Microwave Power*, 1972, **7**:117–137.
21. Bazant, Z. and Najir, L.J., Nonlinear water diffusion in nonsaturated concrete. *Materiaux et constructions*, 1972, **5**(25):3–20.

22. Bazant, Z.P. and Kaplan, M.F., *Concrete at High Temperatures: Material Properties and Mathematical Models*. Harlow, Essex, UK: Longman Group, 1996.
23. Akbarnezhad, A. and Ong, K.C.G., Thermal stress and pore pressure development in microwave heated concrete. *Computers and Concrete*, 2011, 8(4):425–443.
24. Rei, R.C., Prauznitz J.M., and Poling, R.E., *The Properties of Gases and Liquids*. 4th ed. New York: McGraw Hill, 1987.
25. Bazant, Z.P. and Thonguthai, W., Pore pressure and drying of concrete at high temperature. *Journal of the Engineering Mechanics Division*, 1978, 104(5):1059–1079.
26. Elwood, J.M., A critical review of epidemiological studies of radiofrequency exposure and human cancers. *Environmental Health Perspectives*, 1999, Suppl. 1:155–168.

Chapter 2

Microwave-assisted accelerated curing of precast concrete

2.1 BACKGROUND

Precast concrete is defined as concrete that has been mixed, cast, and cured at a location that is not its final destination. As opposed to *in situ* concrete, which is cast and cured on site, precast concrete is produced by casting in reusable forms and curing in a controlled environment off site (Figure 2.1). Traveling distances between the precast site and the installation site may be a few metres when on-site prefabrication is used to distances of tens or even hundreds of kilometres between precast yard and construction site.

Because of the many economic and environmental advantages, concrete precasting is widely used in modern construction. Prefabrication of concrete components off site generally leads to improved quality because of better control over the production process. In addition, precasting has many other advantages in terms of enhanced construction safety as the production processes take place in a controlled environment with overhead cranes and generally at ground level. Precasting can capitalise savings derived from the high reuse rate of formwork in terms of number of components cast and rate of demoulding. Other benefits include faster completion and better quality of finishes.

Precast concrete has many applications, as both structural and architectural components. Almost every structural component can be precast off site to some degree. Precast architectural panels are commonly used as building facades, including as exterior walls and sound and even blast proofing. It may be noted that the majority of storm water drainage, water and sewage pipes and conduits, and tunnels make use of precast concrete units. In practice, much attention has been placed on improving the efficiency of the precast production process. One major factor with an impact on efficiency is the turnaround time associated with casting and demoulding operations. Achieving an adequate strength level as quickly as possible, sufficient for demoulding and handling operations, is key. A higher early strength development rate can lead to considerable economic and environmental benefits. The strength development rate of concrete is influenced

Figure 2.1 A precast concrete structure.

by a number of parameters, including water-to-cement ratio (w/c), cement type, quantity of cement, and the curing method used [1].

With this in mind, researchers for many years have investigated various techniques to accelerate curing of concrete to achieve higher early strengths. The two most commonly used techniques to accelerate curing of concrete are the use of special concrete mix constituents and curing at elevated temperatures. The use of such techniques, however, has a number of associated drawbacks and limitations. For instance, the use of admixtures and rapid-hardening cements may cause long-term durability problems that are as yet not fully understood. In addition, the rapid strength development rate achieved by curing at elevated temperatures may not be sufficient to enable shortening turnaround time to less than 24 hours. Curing at elevated temperatures has also been reported to result in a lowering of long-term strength potential as well as production of a nonuniform distribution of hydration products, with inherent weak zones within the cement matrix. Moreover, conventional heating methods are known to be low in energy efficiency and when applied to larger or thicker sections generally lead to nonuniform heating, causing undesirable temperature gradients to develop between the heated exterior and the cooler interior. The last may result in the development of locked-in differential thermal stresses, leading to microcracks in concrete. In addition, such relatively high-energy-consuming techniques are not environmentally friendly.

This chapter presents a technique based on microwave heating technology that is more efficient and consumes less energy to accelerate curing of concrete. Microwave energy has the potential to be harnessed as a replacement for conventional heating-based curing techniques. The deep penetration of microwaves at specific industrial, scientific, and medical (ISM) frequencies allows for significantly more uniform heating of concrete than achievable using conventional heating methods. This should result in more uniform hydration of the cement paste and a significant reduction in the development of microcracks within the concrete being cured because of the reduction in differential thermal stresses. Unlike concrete that is cured using conventional elevated temperature curing methods, concrete cured using microwaves is expected to achieve similar or higher long-term strengths when compared to concrete cured conventionally [2]. However, optimising the microwave-curing process to achieve the necessary uniform heating throughout the concrete being cured at a specific elevated temperature requires a good understanding of the microwave curing phenomenon and the effects of various influencing parameters. This chapter starts by reviewing some of the basic concepts of strength development in concrete required for a better understanding of the effects of the microwave-curing process. The working principles of a number of conventional curing methods are also reviewed for comparisons. The chapter then introduces the microwave-curing process and its working principles. Available literature on microwave curing is reviewed and discussed.

2.2 HYDRATION OF CEMENT AND STRENGTH DEVELOPMENT

The accelerated strength development in concrete mixtures is a direct outcome of increasing the hydration rate of cementitious materials present. Therefore, understanding the cement hydration phenomenon is essential for comprehending the phenomena leading to an increased strength development rate in concrete cured using accelerated curing methods. The chemical reactions describing the hydration of cement have been identified by investigating the hydration of the individual cement constituents. Although there are a number of cementitious materials that may be used as ordinary Portland cement (OPC) replacements, in the following section, focus is placed on the main constituents of OPC and the general effects they have on the hydration process. The hydration process is then discussed in more detail.

2.2.1 Constituents of ordinary Portland cement and their contribution to strength development

Table 2.1 shows the typical chemical composition of OPC. It should be noted that the quantities do not add to 100% because of the impurities

Table 2.1 Composition of ordinary Portland cement

Chemical name	Chemical formula	Short notation	Weight percentage
Tricalcium silicate	$3CaO.SiO_2$	C_3S	55
Dicalcium silicate	$2CaO.SiO_2$	C_2S	18
Tricalcium aluminate	$3CaO.Al_2O_3$	C_3A	10
Tetracalcium aluminoferrite	$3CaO.Al_2O_3.Fe_2O_3$	C_4AF	8
Calcium sulfate dehydrate (gypsum)	$CaSO_4.2H_2O$	CSH_2	6

usually present. Calcium silicates are the main component of cement, accounting for almost three-fourths of its composition. In fact, calcium silicates are the components responsible for the main cementing properties. The presence of tricalcium silicate accounts for the main difference between the early generation of cements and the OPC produced currently.

When mixed with water, the cement constituents undergo a series of chemical reactions that lead to the hardening of the mix. The reaction of cement constituents with water is referred to as hydration, and the products of this reaction are referred to as hydration products. Characteristics of hydration of cement constituents, including the reaction rate, heat of reaction, and contribution to the resulting strength development are summarised in Table 2.2. As shown, C_3A and C_3S are the most reactive constituents, whereas C_2S has a considerably slower hydration rate. However, the rate of hydration does not necessarily have a direct influence on the strength development rate. In fact, as shown in Figure 2.2, the calcium silicates are responsible for most of the strength development of Portland cement. C_3S is responsible for most of the early strength developing in the first 4 weeks, and both C_3S and C_2S contribute almost equally to the long-term strength of concrete. All Portland cement hydration reactions liberate heat and thus are exothermic. As a result, concrete heats up continuously internally during the hydration process. The heat liberation of a typical Portland cement concrete is estimated to be about 265 J/g and 473 J/g after 3 days and 1 year, respectively.

Table 2.2 Hydration characteristics of OPC constituents

Cement constituent	Reaction rate	Reaction heat liberation	Contribution to cement strength
C_3S	Moderate	Moderate	High
C_2S	Slow	Low	Initially low but high later
$C_3A + CSH_2$	Fast	Very high	Low
$C_4AF + CSH_2$	Moderate	Moderate	Low

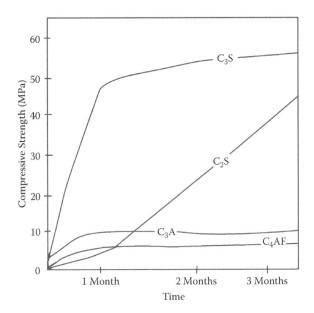

Figure 2.2 Strength development of pure cement compounds.

Table 2.3 Typical Composition of ASTM types I to V
Portland cement

Cement constituents	I	II	III	IV	V
C_3S	55	55	55	42	55
C_2S	18	19	17	32	22
C_3A	10	6	10	4	4
C_4AF	8	11	8	15	12
CSH_2	6	5	6	4	4
Fineness (m²/kg)	365	375	550	340	380

Because of the differences between properties of the various constituents of cement, it is possible to modify the properties of cement by modifying its composition. This is commonly done in many countries. In the United States, five types of cement have been recognized by the American Society for Testing and Materials (ASTM). These types have been developed by varying the proportions of different compounds and the degree of fineness of cement particles. Typical compositions of these cement types are summarised in Table 2.3.

When no special properties are required, type I cement is the most commonly used ASTM cement in construction. However, in applications such as precast concrete, for which a faster strength development rate is required, as well as when concreting in low temperatures, type III cement is the most suitable choice of Portland cement. The increased strength development rate in type III Portland cement is achieved by increasing the content of C_3S and grinding cement to finer particles. The latter strategy is deemed to be more effective because the higher surface area of the smaller cement particles increases the surface area that will come in contact with water. The 24-hour strength development rate of a type III Portland cement is expected to be almost twice that of a type I cement. Type IV cement has been developed mainly to deal with concerns associated with the high heat liberation of type III cement, which could be problematic in mass concrete and when casting large concrete components, for which subsequent cooling of the concrete may cause cracking. Temperature increases as high as 30°C have been reported in the case of mass concrete. In type IV cement, the amount of liberated heat has been reduced by limiting the relative amounts of C_3S and C_3A, which are responsible for most of the heat liberation in early ages. Type V Portland cement was developed to deal with the problem of sulfate attack, deterioration of concrete exposed to water or soils containing sulphates at early ages. In type V cement, this is achieved by lowering the C_3A content to below 5% as sulphate attack only involves interactions between the hydration products of C_3A. Good but relatively less sulphate resistance than type V cement is also achievable by using type IV cement, which has about 8% C_3A. However, because the lower amount of C_3A can result in a considerable decrease in the early strength development rate, type II cement was developed for this purpose by increasing the C_3S content. The compressive strength development trends of different ASTM cements are compared in Figure 2.3.

The ASTM cements described have been developed to control specific aspects of the hydration of cement by varying the proportion of different constituents. However, because the constituents used are similar, the properties of hardened mortar or concrete made with any ASTM cement tend to be similar. Therefore, although the initial rate of strength development is different, the ultimate strength achievable using different cement types tends to be within a similar range, although the lower initial strength development rate usually leads to slightly higher ultimate strength (Figure 2.3).

2.2.2 Hydration of cement

Understanding the hydration process of cement constituents can be beneficial in understanding the effects of the curing processes on the strength development rate in concrete. As discussed, C_3S and C_2S are the main constituents contributing to the development of strength in concrete. Therefore, in this section, focus is placed on the hydration of these compounds. The

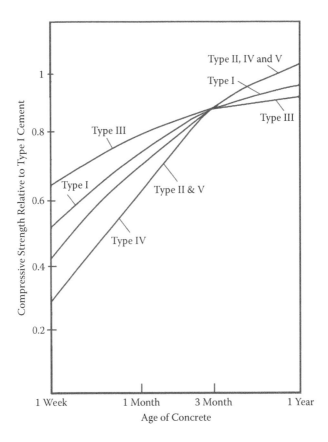

Figure 2.3 Relative compressive strength development trend of concrete made with the same aggregates but with different types of ASTM Portland cements.

hydration reaction for C_3S and C_2S are relatively similar and differ only in the amount of calcium hydroxide (CH) created:

$$2C_3S + 11\ H \rightarrow C_3S_3H_8 + 3\ CH \tag{2.1}$$

$$2C_2S + 9\ H \rightarrow C_3S_3H_8 + CH \tag{2.2}$$

Calcium silicate hydroxide ($C_3S_3H_8$) is the main product of hydration of C_3S and C_2S. $C_3S_3H_8$ is usually referred to using the short format of C-S-H. C-S-H is a poorly crystalline material formed at very small particle sizes in the range of less than 1 μm in the three volumetric dimensions. Both the reactions release heat during the process, which will lead to an increase in the temperature of the concrete. The sequence of the reactions of C_3S and

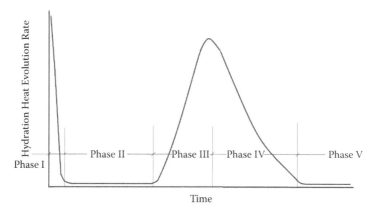

Figure 2.4 Different phases of hydration of C_3S.

C_2S can therefore be characterised by quantifying the heat liberation rate as shown in Figure 2.4. With this method, the hydration process can be broken down into five distinct phases. In phase I, which starts with the mixing of cement and water, a period of rapid heat liberation begins that lasts for about 15 minutes. This phase is followed by a phase of relative inactivity known as the induction period (phase II). The occurrence of phase II is the main reason leading to the plastic state of cement for several hours after mixing with water. The initial set starts at the end of phase II (after 2 to 4 hours) when C_3S starts to react again. The rate of hydration increases rapidly until it reaches its maximum at the end of phase III. At this point (after about 8 hours), the final set has been reached and early hardening has started. This is followed by phase IV, in which the hydration rate slows until it becomes steady in phase V.

C_2S undergoes a relatively similar hydration process but at a relatively slower rate as it is less reactive than C_3S. The underlying phenomena leading to the development of the different stages described have been studied extensively and are well known. A more detailed discussion of the hydration stages, however, is not in the scope of this book and can be found in most concrete technology books.

2.3 CURING OF CONCRETE

2.3.1 Standard curing at ambient temperatures

Many concrete structures fail because the design target strengths are not achieved in practice. In addition, miscalculations in the estimation of the strength development rate or inability to achieve the rate in practice may cause problems when the forms are removed. Achieving the planned properties of concrete, including durability properties, is also essential to ensure the long-term durability of the concrete structure. As discussed in the previous section, the strength and durability of concrete are affected considerably by the degree and rate of hydration. A sufficient level of hydration is required to reduce the porosity of concrete so that the desired strength and durability of concrete are attainable. It is well known that achieving a sufficient level of hydration requires the availability of an adequate supply of water. The surface of concrete is particularly susceptible to incomplete hydration because it dries faster than the interior. With this in mind, curing processes were initially developed to ensure the availability of sufficient water for the complete hydration of cement. The results of numerous studies have shown that curing is a crucial stage in the concrete life cycle to ensure that the optimal properties of concrete are developed. The strength, durability, abrasion resistance, water tightness, volume stability, and freeze-thaw resistance of concrete all increase with adequate curing. Because of such significant effects, the curing of concrete is sometimes considered as a separately paid professional service in especial circumstances.

It is theoretically impossible to achieve fully complete hydration of all cement present in the concrete. This is mainly because C-S-H produced during hydration tends to cover some of the larger cement grains, preventing further hydration reaction. With this in mind, the objective of conventional standard curing is to achieve as much hydration as possible at a reasonable cost. In theory, at w/c ratios equal to or greater than 0.42, there should be sufficient water for achieving complete hydration of all the cement present in the concrete. However, in practice, a considerable amount of the water present in the concrete is evaporated or absorbed by aggregates or forms. Hydration of cement does not require a fully saturated condition and may occur at a relative humidity that is below 100%. This is because cement can use the water held through surface tension in the larger capillary pores. However, it is well known that the rate of hydration tends to be slower at lower relative humidity and ceases when the relative humidity falls below 80%. The water present in concrete is first used for the localised hydration of the adjacent cement particles; thus, the areas with a faster hydration rate tend to experience lower relative humidity with time. If the concrete is maintained at a saturated condition by providing additional water, the water could flow toward the areas with higher consumption rates. However, when the concrete is not maintained at a saturated condition, the flowability of water in the capillary pore system decreases with the gradual decrease in the water content of capillary pores, leading to a considerable decrease in the flow rate of water toward the areas in need of more water. Therefore, even if a concrete is sealed against moisture loss, because of the reduction in the relative humidity with hydration and the subsequent reduction in the ability of the water present in the capillary pores to flow, the rate of hydration and strength development will be slower than for a concrete that is continuously

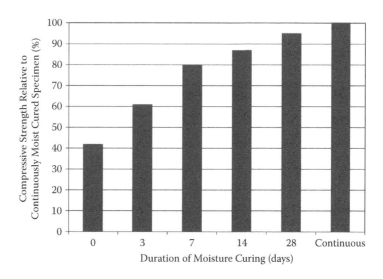

Figure 2.5 Effect of duration of water curing on the long-term strength of concrete.

moist cured. The provision of additional water during the curing process is especially important in concretes with low w/c ratios as the hydration may cease when the relative humidity falls below 80%. Provision of sufficient water for thorough moist curing is sometimes difficult in concretes with w/c ratios lower than 0.3 because of the low permeability of such concretes.

The strength development rate of concrete at ambient conditions varies with the moist-curing scheme applied. The strength development rate slows considerably when the moist-curing period ends. Moreover, continuous moist curing considerably increases the 28-day strength as well as the long-term strength development of the concrete. The ultimate compressive strength of concrete also varies considerably with the duration of moist curing. Figure 2.5 shows that the long-term compressive strength of a concrete that has not been moist cured can be as low as 40% of that of a continuously moist-cured concrete. As shown, this can increase to about 85% and 95% by increasing the duration of moist curing from 3 days to 14 days and 28 days, respectively.

The curing process also affects the durability of concrete considerably. Two of the most important parameters influencing the durability of concrete are permeability and absorptivity, which are related to the porosity of concrete and interconnectivity of capillary pores. Adequate water curing to ensure continuation of hydration can result in partial or complete filling of some of the capillary pores with time, thereby reducing the permeability of the concrete. A decrease in permeability can result in an increase in the durability of concrete by preventing the ingress and transport of harmful substances into the concrete. Figure 2.6 shows that the coefficient

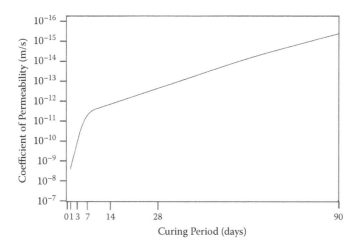

Figure 2.6 Effect of duration of water curing on the permeability of cement paste with a water-to-cement ratio of 0.5.

of permeability of cement paste at a particular w/c ratio reduces considerably with an increase in the curing duration. The effect of curing duration is especially considerable at early ages. It should be noted that the w/c ratio is the single parameter with the largest influence on durability of concrete. Generally, the porosity of concrete decreases with a decrease in the w/c ratio.

2.3.2 Effect of temperature

Temperature is an important parameter affecting the curing rate, especially the curing duration required to achieve a particular strength level. Figure 2.7 shows the variation in the 3-day and 28-day strengths of concrete with an increase in the curing temperature. As shown, an increase in the curing temperature can result in a considerable increase in the early strength development rate of concrete. However, although the increased strength development rate at higher curing temperatures is beneficial, there is a downside to applying higher curing temperatures. As shown in Figure 2.7, the increase in the early strength development usually leads to a decrease in the long-term and ultimate strength gains of the concrete that is not desirable.

The increased strength development rate of concrete at elevated temperatures is because of the well-known increase in the cement hydration rate at higher temperatures. The temperature sensitivity of the hydration process is highest in phase III of the hydration process. The temperature sensitivity tends to decrease significantly in phase V as the hydration process becomes more diffusion controlled rather than chemically controlled.

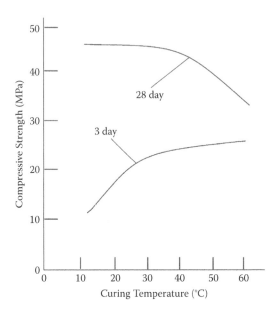

Figure 2.7 The 3-day and 28-day strength of concrete cured at various constant curing temperatures.

Despite minor changes in the composition of C-S-H, the stoichiometry of hydration remains the same up to about 100°C. However, although the reasons for the increased strength development rate of concrete at elevated temperatures are mostly known, the reasons behind the decreased long-term strength of the concrete when cured at elevated temperatures are not as clear. This is partially because an increase in temperature of up to 45°C is considered to have no or negligible effects on the physical or chemical structure of the hydration products. It has been suggested by some studies that the negative effects of elevated temperatures on the long-term strength of concrete stem from the nonuniform distribution of hydrated cements within the paste at elevated temperatures, which results in weak parts present in the matrix [1].

The inverse relationship between the early and ultimate strength of concrete is usually considered as a general rule; i.e., the higher the initial strength of concrete, the lower the long-term strength will be. The detrimental effects of higher curing temperature and higher initial strength development rate on the long-term strength of concrete are also seen even when the concrete is initially cured at elevated temperatures and then kept at ambient temperatures (Figure 2.8). A similar inverse relationship between early and long-term strength of concrete is also observed when curing the concrete at lower temperatures. This has been shown to result in a decrease

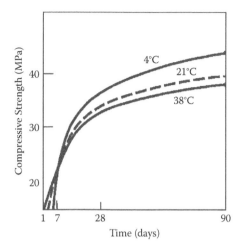

Figure 2.8 Compressive strength of concrete cured initially at different temperatures followed by curing continuously at 21°C. (Reprinted, with permission, from *Proceedings-American Society for Testing and Materials*, 1951, copyright ASTM International, 100 Barr Harbor Drive, West Conshohocken, PA 19428.)

in the early strength but an increase in the long-term strength of concrete. Temperatures below 10°C are generally undesirable for the development of early strength of concrete. The development of early strength is considerably retarded at temperatures below 4°C. The decrease in the early strength development rate continues with a further decrease in temperature, down to –10°C, at which little or no strength development will occur.

2.3.3 Standard curing methods

The curing process after placement of concrete in the formworks is traditionally performed using one of the following two methods:

1. *Water curing*, in which measures are taken to supply additional water to compensate for the loss of moisture and to maintain sufficient mixing water in the concrete during the early hardening days.
2. *Sealed curing*, in which measures are taken to reduce the loss of moisture from the surface of concrete.

2.3.3.1 Water curing

In water curing, water is supplied directly to the concrete using various methods, including ponding, spraying, or covering the concrete surface with saturated coverings. The temperature of water used for curing should

Figure 2.9 Ponding of a concrete floor. (Courtesy of Alec Johnson, Nong Khai Design Co., Ltd., 1070/1 Prajak Road, Tambon Nai Muang, Amphoe Muang, Changwat Nong Khai, 43000 Thailand.)

not be more than 5°C lower than that of the concrete to avoid development and growth of thermal cracks. Ponding is mostly used in the curing of large horizontal surfaces such as concrete floors and pavements. In this method, a dam or dike is erected around the edges of the horizontal concrete element to maintain the ponding water, which is added later (Figure 2.9). Ponding is an effective curing method for preventing moisture loss from concrete and maintaining a uniform temperature within the concrete. Ponding is a fast, effective, and inexpensive curing method when water is plentiful and when there is a ready supply of dam materials such as clay. Prior to applying the ponding method, it should be ensured that ponding does not disrupt subsequent construction operations. Because of the considerable labour and level of supervision requirements, ponding is usually used only for small jobs.

Spraying is an alternative water curing method that is also commonly used when water is easily accessible. Spraying is especially effective when the ambient temperature is well above the freezing temperature and when ambient humidity is low. This method is applicable for both horizontal and vertical surfaces. In water spraying, a fine spray such as that used by most sprinklers is used to continuously spray water onto the surface and thereby increase the relative humidity of the air over the concrete surface to slow evaporation from the surface. Similar to other wet curing methods, it is crucial that the sprinklers are able to maintain a continuously wet concrete surface during curing. This does not require the continuous operation of sprinklers, and timers may be used based on the weather conditions and the estimated rate of evaporation. When spraying is carried out intermittently, burlaps or other similar materials should be used to prevent the drying out of the exposed concrete surface. This is because alternate cycles of drying and wetting can cause surface cracking. The relatively high water supply requirement is the main disadvantage of the spraying method. Closed-system sprinklers capable of collecting and recycling the water used can save a lot of water and are therefore recommended. The effectiveness of wet curing using sprinklers is highly affected by windy conditions. In such conditions, adequate supervision is required to ensure that the entire exposed surface of the concrete is maintained moist without alternate wetting and drying of any part of the concrete being cured.

Covering the concrete surface with saturated coverings such as burlap, cotton mats, rugs, or any other moisture-retaining materials is another common wet curing method that can be applied to both horizontal and vertical concrete surfaces. The requirements for burlap and burlap-polyethylene sheeting suitable for curing are specified, respectively, in the AASHTO (American Association of State Highway and Transportation Officials) M182 and ASTM C171 standards. The coverings should be free of any substance that may be harmful to concrete, and new coverings should be rinsed with water to remove soluble substances. Before placement of coverings on the surface of concrete, it should be ensured that concrete has hardened sufficiently such that any damage to the surface of concrete is unlikely. Also, the coverings are to be supplied with sufficient water to maintain the required water content in the concrete during curing. Wet coverings of sawdust, sand, and earth are also effective for the curing of concrete and are used in small jobs. In such cases, a layer approximately 50 mm thick of wet covering materials should be evenly distributed over the prewetted surface of the concrete and maintained wet continuously. The main disadvantage of using wet sawdust, sand, or earth coverings is the possibility of staining the concrete surface.

In all these wet curing methods, it is important that the supply of water should start at least 1 hour before the initial set, that is, before the end of the induction period. This is because maintaining a continuous supply of water, which facilitates the ingress and transport of free water to the large capillary pores present in the concrete, is essential for the free water to reach the unhydrated cement particles. A delay in the provision of sufficient free water may result in the development of menisci in capillary pores, which render the penetration of water into the capillary pores more difficult. The importance of this issue increases with a decrease in the w/c ratio of the concrete being cured.

2.3.3.2 Sealed curing

Sealed curing methods are based on preventing or minimising moisture loss from the exposed surfaces of concrete. Because of the ease of application and low labour requirements, sealed curing using plastic sheeting, curing membranes, and waterproof papers are preferred over some of the traditional wet curing methods mentioned previously. When applying such sealed curing methods, the absorption of water by formwork and loss of moisture through forms should be also minimised. This can be achieved by proper selection of materials with minimal water absorption for fabrication of formworks and appropriate design of formworks against leakage. Such coverings should be used after the concrete surface has been wetted sufficiently and has hardened sufficiently.

Plastic sheets, or other similar materials, may be used as effective barriers against water loss (Figure 2.10). To provide sufficient sealing effect, such sheets should be secured in place and protected from damage. When used in the curing of flat concrete surfaces, such as concrete pavements, such sealing sheets should extend beyond the edges of the slab for at least twice the slab thickness or be turned down over the edges of the slab and sealed. Wrinkling of the covering sheets should be avoided to reduce the risk of hydration staining. Curing with polyethylene sheets may cause patchy discolouration of the concrete surface, especially if the concrete contains calcium chloride. Such discolouration tends to be more prevalent when the covering is wrinkled. This is especially important in the curing of concrete

Figure 2.10 Polyethylene plastic films acting as an effective moisture barrier for curing the concrete.

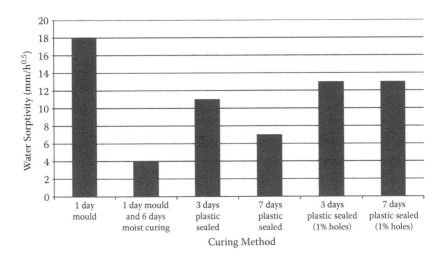

Figure 2.11 Effectiveness of plastic sheeting compared to standard curing.

with decorative finishes, for which uniformity of colour is required. For vertical concrete elements, the sheets should be wrapped around the element and held in place by adhesive or ducting tapes to limit moisture loss. An adequate thickness of the sealing sheet is important to prevent excessive moisture loss, and a thickness of at least 0.01 mm has been recommended by ASTM C171. In addition, any damage to the sheets during use should be prevented. The presence of even small holes could reduce curing efficiency considerably (Figure 2.11).

The plastic sheets or other covering materials used in sealed curing can be transparent or coloured. The colour of the coverings can considerably affect the curing process and should be selected by considering the ambient conditions. For instance, in hot weather, black plastic coverings can lead to unacceptably high concrete surface temperatures through higher heat absorption. In such cases, lightly coloured sheets are preferable. On the contrary, the use of black coloured sheets can be highly beneficial for accelerating the rate of the concrete strength gain in cold weather. The effectiveness of sealed curing depends very much on the integrity of the sealing material and the thickness of the covering. When multiple sheets are used, proper overlaps and joints should be used for effective sealing against moisture loss.

Another relatively similar sealed curing method is curing using impervious paper. This paper consists of two sheets of kraft paper cemented together by a bituminous adhesive with fibre reinforcement. The requirements of impervious papers for the curing of concrete are specified in ASTM C171 and AASHTO M171 standards. Edges of adjacent paper sheets are

Figure 2.12 Spray application of liquid membrane-forming curing compound onto the concrete surface. (Courtesy of Derek Tibbits, Termguard, Perth Head Office, 84 Welshpool Road, Welshpool, WA 6106, Australia.)

recommended to be overlapped by about 150 mm and sealed with wood planks, sand, or glue. One of the advantages of using impervious papers is reusability. Any tears or holes and other such damage can be easily repaired using curing-paper patches prior to reuse.

Besides the plastic and paper sheets, curing membrane-forming compounds are also commonly used as covering materials for sealing the exposed surfaces of concrete to prevent moisture loss in the initial stages. Similar to curing using plastic and impervious paper sheeting, an important advantage of this method is that periodic additions of water are not required. Membrane-making compounds dissolved in a volatile solvent or water are usually sprayed directly onto the concrete surfaces. An almost-impermeable layer remains on the concrete surface after evaporation of the solvent and seals the concrete (Figure 2.12). The properties and applications of membrane-making materials for the curing of concrete are described in the Australian Standard (AS) 3799 standard. Membrane sealing is one the most widely used methods for extending the curing of concrete after an initial water curing stage. However, curing membranes should be avoided when the surface of concrete is to be painted or overlaid with a topping. This is because the membrane will weaken the bond between the finishes and the concrete surface. Most curing compounds are not compatible with the adhesives used to lay flooring materials. Membrane-forming compounds can be translucent or white pigmented. A dye is usually added to the translucent compounds to enable checking for full coverage of the concrete surface by the compound. The dye colour usually fades soon after formation of the membrane.

2.4 ACCELERATED CURING OF PRECAST CONCRETE

As discussed, shortening the curing duration while ensuring the quality of the curing process can lead to considerable economic and environmental advantages in precast concrete construction. A shorter curing duration will result in faster demoulding of formwork, which leads to shorter turnover time, a higher production rate, and fewer formworks required. In addition, faster production and thus earlier delivery of precast components to the construction site can shorten the construction time and therefore the completion time of the project. Earlier project completion can translate to considerable economic advantages for the project owners. Ensuring adequate curing using conventional techniques such as wet curing or sealing takes a longer time and requires proper supervision. This can be factored in during the whole production process of precast concrete, but there are potential economic gains to be reaped if curing of the precast concrete components can be shortened through accelerating strength gain in concrete used in precasting operations.

Based on the previous discussions on the factors affecting hydration rate, two main approaches are generally adopted to achieve high early strength in concrete. In the first approach, the process is accelerated by changing the curing conditions. In this case, the main factor affecting the rate of hydration is the curing temperature. In the second approach, the composition of the cement is modified to increase the early strength development. In this approach, the increased strength development rate is achieved using special cements or mineral and chemical admixtures. In practice, a combination of both techniques is usually applied to achieve an effective and economical accelerated curing regime for the concrete used.

In the following sections, we first review some of the traditional accelerated concrete-curing methods used in the precast concrete industry. The methodology as well as the advantages and disadvantages of each method are discussed. We then describe a microwave-assisted concrete curing method that may be used to reduce curing durations to levels not easily achievable using the other more conventional accelerated curing regimes. The working principles of the microwave-curing method as well as its advantages and disadvantages are discussed in detail.

2.4.1 Physical processes for accelerated curing

The relationship between the compressive strength development rate of concrete and temperature was discussed in Section 2.3.2. We observed that an increase in the curing temperature usually results in an increase in the strength development rate at early ages. However, the increased strength development rate occurs only up to a certain temperature, beyond which any additional increase in temperature is not only less effective but may also have detrimental effects on the concrete properties. A variety of methods for accelerated curing of concrete at elevated temperatures have been investigated and used over the past few decades. These methods include convection heating by circulation of a hot liquid (water or oil) through the formwork, electric resistance heating, low- and high-pressure steam curing, and microwave curing. One of the drawbacks associated with most of these elevated temperature curing methods is the increase in the rate of the humidity loss from concrete at elevated temperatures, which can in turn lead to severe shrinkage and cracking problems. Therefore, any method for the curing of concrete at elevated temperatures should also ensure adequate humidity to prevent excessive moisture loss.

Four main parameters should be considered generally when designing an elevated-temperature concrete-curing process: rate of temperature rise, maximum temperature, rate of heating, and uniformity of heating. Controlling the rate and uniformity of heating are highly important to minimise the development of differential thermal stresses in the concrete. Curing at elevated temperatures using the majority of conventional heating methods may result in the development of differential thermal stresses as a result of the rapid increase in the temperature as well as the non-uniform heating of the concrete elements. Such differential thermal stresses negatively affect the properties of the cured concrete through introducing new microcracks or increasing the growth rate of the pre-existing cracks. It is generally recommended to restrict the maximum curing temperature of concrete to 60°C to 70°C. The maximum temperature limits in concrete curing at elevated temperatures are mainly caused by the well-known decrease in the long-term strength of concrete with an increase in its early strength development rate. In the following, various traditional elevated-temperature curing methods are reviewed. The microwave-curing method is discussed in detail later in this chapter.

2.4.1.1 Accelerated curing using convection and conduction

Curing at elevated temperatures is traditionally performed using some form of conduction/convection heating-based methods. One of the methods commonly used is to increase the temperature of the forms by pumping hot water or hot oil through them or by heating them electrically [3]. The temperature of the concrete placed in the forms being heated gradually increases through conductive heat transfer. Proper insulation of formwork is highly important to improve the energy efficiency of this method. In addition, similar to any other method based on curing at elevated temperatures, ensuring sufficient humidity to compensate for the humidity loss during high-temperature curing is important to maximise the benefits derived.

2.4.1.2 Accelerated curing using electrical heating

Generally, two different methods have been suggested for the curing of concrete using electrical heating. In the first method, the electrical resistance of the embedded reinforcing bars or additional elements, such as special coils of wire embedded in the concrete, is used to heat the concrete [4]. When electrical current flows through the embedded wires or reinforcing bars, the electrical resistance of these conductors results in the generation of heat increasing the temperature and thus accelerating curing of the concrete component.

The second method used commonly in the electrical curing of concrete is referred to as direct electrical curing. This method capitalises on the inherent electrical resistivity of concrete, which is estimated to be in the range of 100 ohm/metre for typical concretes to heat the concrete by passing an alternating current through it [4]. In this method, the use of additional wires or formworks as the heating medium is unnecessary. It is believed that, compared to the first method, direct electrical curing may result in more uniform heating of the concrete.

2.4.1.3 Accelerated curing using low-pressure steam

In this method, accelerated strength development of concrete is achieved using low-pressure steam, which simultaneously elevates the temperature and provides concrete with additional moisture to compensate for moisture loss. The temperature elevation could be achieved either as a direct result of steam injection or through an alternate heating method followed by an increase in the humidity through steam injection. To minimise the loss of heat and moisture, steam curing is usually performed in an enclosed environment.

Low-pressure steam curing is especially appealing in dry climates when it is imperative to control the moisture loss of concrete. The operation cycle of typical low-pressure steam curing is shown in Figure 2.13. The curing process usually starts with an initial delay period of about 3 to 5 hours, which is necessary for the concrete to set. This is then followed by a heating period, beginning with a gradual increase in the temperature at a rate of about 10°C to 20°C/hour until the temperature reaches its target maximum level, usually ranging from 60°C to 70°C. Next, the concrete is subjected to steaming at the maximum curing temperature for a period of about 6 to 12 hours. Temperatures above 70°C should be avoided as it is uneconomic and may cause heat-induced delayed expansion and undue reduction in the ultimate strength. The steaming period is followed by a cooling period. Concrete should still be covered with steam hoods or traps during the cooling period.

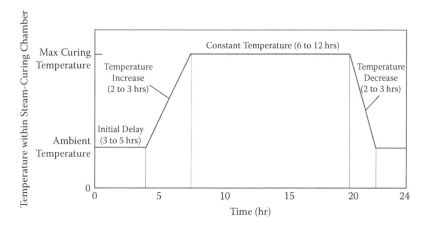

Figure 2.13 Typical low-pressure steam-curing cycle.

It is sometimes necessary to release some of the boundary constraints caused by the forms, prestressing tendons, and so on to prevent any damage caused by the development of differential thermal stresses during the cooling period. Finally, during the last stage, which is referred to as the exposure period, the steam hoods and traps are removed, and the concrete is exposed to ambient conditions [3]. The concrete temperature should be monitored at the exposed ends of the concrete elements. The temperature of the air may not be a good representative of the internal temperature of concrete because the internal temperature may also rise because of the heat generated by hydration.

The effects of low-pressure steam curing on various properties of concrete have been widely studied according to the available literature. The results of previous studies generally showed that the combined effects of the elevated temperature and moist curing provided by low-pressure steam curing can lead to the development of high early strengths in concrete. Other reported benefits of steam curing include reduced drying shrinkage and creep compared to moist-cured concretes [5]. The low-pressure steam curing method has also been reported to result in slight reductions in the ultimate strength of concrete. However, such reductions are insignificant and do not offset the benefits of this method. The ultimate strength of concrete is influenced more by the long-term availability of moisture than the low-pressure curing process itself.

2.4.1.4 Accelerated curing using high-pressure steam

In high-pressure steam curing, also referred to as autoclaving, the increase in the curing temperature and humidity is accompanied by an increase in

Figure 2.14 A typical concrete-curing autoclave.

pressure. Because of the use of high pressures, concrete components being cured have to be placed in an enclosed vessel thus restricting the application of this method to small components (Figure 2.14). One of the main advantages of autoclaving is that with this method extremely low w/c ratios can be used in the mix design. This is because sufficient moisture to complete the hydration process is injected into the concrete components through the high-pressure steam. The concrete components cured using high-pressure steam are usually produced through extrusion machines using no-slump concrete. However, high-pressure steam curing also has a number of disadvantages. It has been reported that, during autoclaving, pretensioned concrete components may experience as much as 20% relaxation, which has to be accounted for during the design phase [3]. Moreover, autoclaving increases the creep, which also leads to a loss of prestressing force. Besides these concerns, there has been no report on the possible detrimental effects of the use of high-pressure steam in curing on the long-term strength of concrete.

2.4.2 Accelerated curing using mineral admixtures

2.4.2.1 Silica fume

Silica fume is an amorphous polymorph of silicon dioxide. This mineral, which is also known as microsilica, is an ultrafine powder and is usually produced as a by-product of silicon and ferrosilicon alloy production. Silica fume particles are spherical, with an average diameter of 150 nm, making

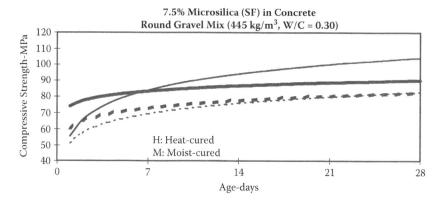

Figure 2.15 Effect of silica fume on concrete strength. (From French, C., Mokhtarzadeh, A., et al., High-strength concrete applications to prestressed bridge girders. *Construction and Building Materials*, 1998, **12**(2–3):105–113. With permission.)

them approximately 100 times smaller than the average cement particles. Because of its high fineness and high silica content, silica fume is a highly reactive pozzolanic material and has been widely used as Portland cement replacements to enhance the properties of concrete.

Silica fume can considerably increase the compressive strength of concrete. This includes an increase in both early and ultimate concrete strengths. Figure 2.15 compares the compressive strength of the concretes made with and without addition of silica fume (7.5% by weight of cement) at two different curing conditions. As shown, the addition of silica fume combined with curing at elevated temperatures may lead to higher compressive strengths. The increase in the compressive strength of concrete with the addition of silica fume is attributed to the improved characteristics at the interfacial transition zone in concrete, leading to an increased aggregate/mortar bond strength as well as improved cohesiveness and decrease in segregation and bleeding of concrete [6]. Such improvements result from the packing effects caused by the addition of very fine silica fume powder to the cement paste and from the pozzolanic reactions between free calcium hydroxide in the paste and silica fume [7]. It should be noted that the positive effect of silica fume on the compressive strength is not limited to the early stages of curing.

In addition to the positive effects on the ultimate strength and strength development rate of concrete, the use of silica fume has other advantages, such as decreasing the permeability of concrete to chloride ion ingress, which reduces the risk of corrosion of reinforcing bars embedded in the concrete. This is especially important when concrete is used in chloride-rich environments, including coastal regions [8]. Silica fume has also been reported to reduce the segregation and bleeding problem in the concrete. This is because of the reduction in the free water in the concrete due to the consumption of a considerable portion of the water necessary to wet the relatively large surface areas of silica fume.

2.4.2.2 Calcium chloride

Calcium in the form of calcium chloride has been traditionally used to accelerate the strength development rate of concrete. However, the addition of calcium chloride to concrete has resulted in numerous problems and therefore has been banned in several countries [9]. The main concern associated with the use of calcium chloride in concrete is the contribution of chlorides to corrosion of reinforcing bars.

The effect of calcium chloride on early strength development in concrete has been confirmed by a number of studies. It has been reported that the addition of 1% (by weight of cement) of calcium chloride may result in up to a 300% increase in the 24-hour strength of concrete [9]. Because of such benefits with regard to acceleration of the early strength gain, the use of calcium chloride in concretes without metal reinforcement is still permitted. The content of calcium chloride has been recommended to be limited to 2% by weight of the cement. In addition, caution has been advised when using calcium chloride in concrete intended for steam curing, concrete containing dissimilar metals, prestressed concrete, concrete susceptible to alkali-aggregate reaction, concrete exposed to sulfates, concrete with embedded aluminum, mass concrete, floor slabs intended to receive dry-shake metallic finishes, and coloured concrete [10]. Despite these concerns, calcium chloride is still a frequently used admixture in many countries, including the United States.

2.4.2.3 Superplasticisers (high-range water reducers)

Superplasticisers, also referred to as high-range water reducers (HRWRs) can contribute considerably to increasing the early strength development of concrete under both normal and accelerated curing conditions. As the name implies, water-reducing admixtures reduce the water demand of the concrete mix. HRWRs act as dispersants to avoid particle aggregation and to improve the flow characteristics of concrete. Therefore, HRWRs can be used to increase the workability of concrete while maintaining the desired target strength or to increase the strength of concrete while maintaining the desired target workability. The majority of water-reducing admixtures used conventionally consist of hydroxylated carboxylic acids, lignosulfonic acids, or processed carbohydrates. Such water-reducing admixtures have been shown to result in up to 10% reduction in the water demand for a particular concrete mix for a given workability target. This class of water-reducing admixtures was followed by a more recently developed group of

HRWR admixtures composed of organic polymers that may easily reduce the mix water requirements up to 20 to 25% while maintaining the desired workability [11]. Unlike calcium chloride, the use of HRWR does not cause durability problems related to corrosion as HRWRs do not contain added chlorides.

Depending on the type of the HRWR used, the mix design, mix sequencing, and admixture dosage should be adjusted to increase the benefits derived from the use of superplasticisers. It has been recommended that a short delay (30 to 60 seconds) between the addition of water to the mix and the addition of HRWR admixture is required to optimise the effects of the HRWRs [11]. Determining the optimal dosage of HRWR is highly important, especially when the aim is a higher early strength development rate. There is always a particular dosage above which any further increase in the dosage of HRWR will not contribute to further increases in the strength development rate and may even have negative effects, including severe retardation.

The maximum size of the aggregates in concrete also affects the effectiveness of the HRWR admixtures. The effect is particularly considerable in concretes with higher mortar content, which are common in precast applications. It has been shown that a decrease in the maximum size of aggregates may lead to an improved transition zone and thus an increase in mortar-aggregate bond strength because of the higher surface area of smaller aggregates. The efficiency of HRWRs can also be affected by the composition of cement. The effect of HRWRs on strength development is more significant in concretes with higher cement content and in concretes made with finer cements or cements with lower C_3A content.

2.5 MICROWAVE CURING OF CONCRETE

A number of commonly used traditional curing methods for accelerating the early strength development rate of concrete were reviewed in the previous sections. The main strategy in most of these methods is to accelerate the hydration process through an increase in temperature while ensuring the availability of sufficient internal moisture. However, one of the major disadvantages of traditional accelerated concrete-curing techniques is nonuniformity in the degree of curing (hydration) when concrete elements being cured are relatively large. This is mainly because of the inherent thermal insulating properties of concrete and the nonuniformity of heating achieved when utilising conventional elevated-temperature curing methods. Because of its relatively poor thermal conductivity, concrete requires some time to transfer the heat from its surface to the interior, which leads to a nonuniform temperature distribution when concrete is heated from the exterior using conventional heating methods.

The nonuniform heating of concrete in traditional accelerated curing methods also results in the nonuniform hydration of cement present in different parts of the concrete undergoing curing. In addition, at the relatively high curing temperatures used in conventional curing methods, the early hydration of the cement constituents, especially C_3S, tends to be too rapid. This leads to the formation of a large amount of very fine C-S-H gel surrounded by unhydrates, which hinders diffusion and further development in strength [2]. The last phenomenon has been considered one of the potential causes of the reduction in ultimate strength of the concrete cured using traditional accelerated curing techniques. Nonuniform heating of the concrete may also result in the development of differential thermal stresses, which negatively affect the long-term properties of concrete by creating new nano- and microcracks and accelerating the growth rate of the existing cracks present in the concrete.

Besides the nonuniformity of heating and its unfavourable consequences, another important disadvantage of traditional accelerated curing methods is their long process durations. Even the shortest traditional accelerated curing processes, such as steam curing, require about 10 hours to complete before demoulding can be carried out. The long process duration of traditional curing methods results in higher processing and supervision costs as well as considerable environmental impacts associated with the relatively high amount of energy consumed. Reducing the curing duration is highly appealing to the precast concrete industry because a shorter curing duration could also mean higher productivity and reduced fabrication and storage area requirements, both reducing the costs of prefabrication [2].

In response to the need for a more efficient and less-time-consuming accelerated curing method, a great deal of attention has been placed on using processes that are more energy efficient, such as microwave heating. The microwave-curing technique has been suggested as a potential technique for revolutionising accelerated concrete curing because of the known unique advantages of microwave heating over conventional heating. When fine-tuned through proper selection of process parameters and appropriate design of the curing equipment, microwave curing eliminates some of the main drawbacks associated with conventional accelerated curing techniques through shortening of the precast production time and minimising nonuniformity of heating. The working principles of the microwave-curing method are discussed in the following sections.

2.5.1 Working principles and development history

All concrete mix constituents, including cement, water, aggregates, and admixtures, are dielectric materials and thus absorb heat when exposed to a microwave field. The electromagnetic loss in dielectric molecules results in the dissipation of microwave energy in the form of heat. Similar to the

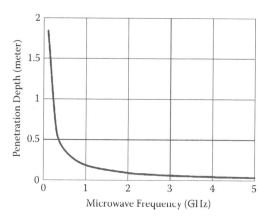

Figure 2.16 Penetration depth of microwave in a saturated concrete at different microwave frequencies.

conventional elevated-temperature curing methods discussed, the elevated
temperature developed as a result of microwave heating can be harnessed
to accelerate the strength development of concrete. The major advantage
of microwave heating over other heating methods is that there is increased
control over the heating process, which can be harnessed to achieve more
uniformity in the heating of the concrete. As discussed in Chapter 1, the
distribution of microwave power and thus the pattern of heating of the
concrete are controlled mainly by microwave frequency and the dielectric
properties of the concrete. The penetration depth of microwaves and therefore the uniformity of power distribution in concrete increase generally
with a decrease in microwave frequency. At low ISM microwave frequencies
(<1 GHz), the penetration depth of microwaves in freshly mixed concrete
easily exceeds the thicknesses of most typical precast concrete elements
(Figure 2.16). Therefore, microwave heating at low ISM frequencies can be
applied to achieve relatively uniform heating patterns for use in the curing
purposes. The penetration depth of microwaves at a particular frequency
tends to increase as concrete dries (Figure 2.17).

Contrary to traditional heating methods, which heat the concrete from
the exterior to the interior, microwaves heat the concrete volumetrically.
Therefore, the rate of microwave heating is not controlled by the thermal
conductivity of the concrete, and a particular uniform temperature profile can be reached in a considerably shorter duration. As a result, microwaves can cure the concrete in a considerably shorter duration vis-à-vis
other more conventional sources of heating. The more uniform heating
achieved using microwaves leads to a more uniform accelerated hydration
of the cement present in the fresh concrete, which improves the mechanical

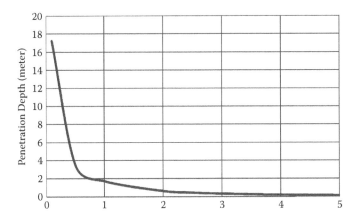

Figure 2.17 Penetration depth of microwave in an air-dried concrete at different microwave frequencies.

properties and durability of the hardened concrete. The improved mechanical and durability properties of microwave-cured concrete are also partially caused by densification of the microstructure of the cured concrete achieved as a result of uniform evaporation of free water at a controlled rate, which can be designed to achieve collapse of the capillary pores of concrete being cured at early ages.

Accelerated curing using microwaves was first proposed by Wu et al. and was applied to cure mortar specimens [2]. In an experimental study conducted by these scientists, the effects of microwave heating on the strength development of mortar specimens was investigated using a basic domestic microwave oven operating at 2.45-GHz frequency. The microwave oven had a maximum power of only 1250 W, which could be regulated at nine levels. The variation in the microwave power in this early-generation domestic microwave oven was achieved through closing of the outlet of the waveguide through the use of a metal plate for different periods of time; i.e., to achieve lower average power, the waveguide had to be closed for a longer period of time. It is now well known that such pulsed microwave transmission is not desirable for curing purposes. The mortar specimens with a sand-to-cement ratio of 2.5 and w/c ratio of 0.44 were subjected to microwave heating immediately after mixing the water. Two different microwave power levels of 1 (~150 W) and 2 (~300 W) and different heating durations of 15 to 120 minutes were investigated to optimise the curing process. The results of these experiments, the first trials of using the microwave-curing method, were promising and marked the beginning of a worldwide research effort on the application of microwave curing to accelerate the early strength development of precast concrete components.

Over the years, further technological improvements in the manufacturing of microwave ovens capable of achieving higher levels of uniformity and the availability of a wider range of microwave frequency sources led to a considerable increase in the feasibility of applying microwave curing in practice. This resulted in a sudden rise in the amount of research to apply microwave curing to mortar and concrete specimens. Hutchison et al. investigated the effects of microwave heating on the degree of reaction and compressive strength of cement [12]. The results of this study were in agreement with the initial observations of Wu et al. in 1987 and confirmed that microwaves can be used as an efficient means for accelerating the strength development of cementitious materials with little, no, or minimal negative effects on longer-term properties. One of the other important findings of this study was related to the effects of microwave-curing duration on the strength development of concrete. It was observed that microwave acts as an accelerator of hydration and reduces the induction period only during the first 24 hours after mixing the cement paste.

In 1994, Bella et al. for the first time applied the microwave energy to cure concrete specimens and reported considerable improvements in terms of the strength development rate compared to those achievable using traditional curing methods [13]. In 1995, Leung and Pheeraphan conducted a comprehensive study to optimise the accelerated microwave-curing process [14]. This study showed that heating rate, power level, w/c ratio, and delay time before the start of microwave curing can considerably affect the outcome of the microwave-curing process. Considerable improvements in the early strength development rate of concrete were reported. In 1997, Leung and Pheeraphan investigated the feasibility of applying the microwave-curing method in practice [15]. Two potential microwave applicator designs for curing of precast concrete blocks and slabs were proposed, as shown in Figures 2.18 and 2.19.

Mak also found microwave heating to be an efficient method for accelerating the early age curing of concrete and discussed some of the requirements of microwave curing to achieve optimal precast concrete performance and process efficiency [16,17]. They also reported on the effects of microwave curing on the performance of structural concrete elements used in bridge construction. The properties considered include thermal response, early age strength development, durability, creep, and shrinkage. It was found that microwave heating results in considerably lower temperature gradients and thus higher heating uniformity than steam curing. In 1999, Sohn and Johnson reported that optimal microwave curing is achieved only at concrete temperatures ranging from 40°C to 60°C [18]. In 2002, Mak et al. showed that microwave-curing cycles of less than 6 hours can provide sufficient strength for removal of formworks and prestressing without affecting the concrete quality negatively [19]. Lee compared the effects of the steam and microwave-curing methods

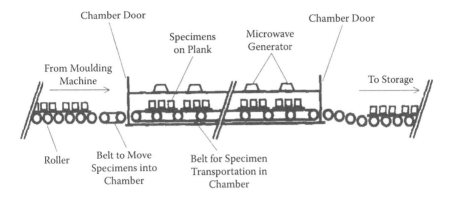

Figure 2.18 A proposed microwave applicator configuration for precast concrete blocks. (From Leung, C.K.Y. and Pheeraphan, T., Microwave curing of Portland-cement concrete—experimental results and feasibility for practical applications. *Construction and Building Materials*, 1995, **9**(2): 67–73. With permission.)

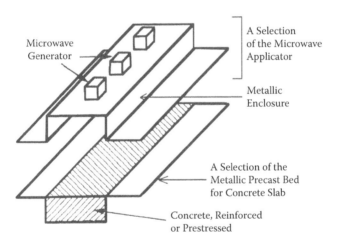

Figure 2.19 A proposed microwave applicator configuration for precast concrete slab. (From Leung, C.K.Y. and Pheeraphan, T., Microwave curing of Portland-cement concrete—experimental results and feasibility for practical applications. *Construction and Building Materials*, 1995, **9**(2): 67–73. With permission.)

on the strength development rate of concrete and found that microwave curing can lead to further improvements while eliminating the concerns about concrete deterioration related to steam-cured concrete [20]. In 2008, Rattanadecho et al. used a continuous microwave drier to accelerate the curing of mortar specimens [21]. The continuous microwave system used had fourteen 800-W magnetrons to achieve a maximum power of 11.2 KW

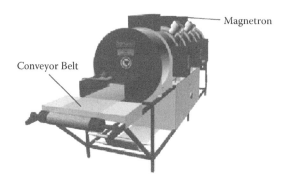

Figure 2.20 A continuous microwave heating system usable for accelerated curing of concrete. (From Rattanadecho, P., Suwannapum, N., et al., Development of compressive strength of cement paste under accelerated curing by using a continuous microwave thermal processor. *Materials Science and Engineering*, 2008, **427**:299–307. With permission.)

(Figure 2.20). Again, the results of this study confirmed that microwave energy can considerably accelerate the early strength gain of the mortar without compromising its long-term strength.

The brief literature review presented generally confirms the capability of microwave curing for accelerating the strength development of concrete. However, achieving optimal properties through microwave curing requires ensuring an optimal curing environment through selection of appropriate heating temperature and curing duration. The heating temperature in microwave curing, however, is a function of various parameters, including the microwave frequency, dielectric properties of concrete being cured, and design of the applicator used. The concepts of estimating the heating temperature based on the microwave power dissipation in concrete and available analytical techniques for estimating the heating temperature were discussed in Chapter 1. The concepts of designing microwave applicators to achieve optimal uniform heating of concrete in accelerated microwave curing are discussed in Chapter 6. Fine-tuning the microwave-curing process to achieve cured concretes with optimal properties requires a thorough understanding of the effects of microwave heating on the various properties of concrete, especially the strength development rate. In the following sections, the effects of microwave curing on various properties of concrete are discussed in detail.

2.5.2 Microwave curing and compressive strength development

Microwave curing has been shown to be highly effective in accelerating the early strength development of cement paste, mortar, and concrete. Early

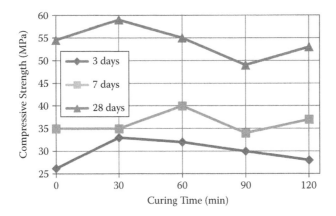

Figure 2.21 Variation in the compressive strength of mortar specimens (w/c = 0.44) with curing time at microwave power of 150 W. (From Wu, et al., Microwave curing technique in concrete manufacture. *Cement and Concrete Research*, 1987, **17**(2):205–210. With permission.)

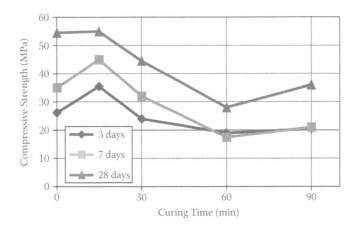

Figure 2.22 Variation in the compressive strength of mortar specimens (w/c = 0.44) with curing time at a microwave power of 300 W. (From Wu, et al., Microwave curing technique in concrete manufacture. *Cement and Concrete Research*, 1987, **17**(2):205–210. With permission.)

studies of microwave curing showed that, unlike steam curing, microwave curing applied for optimal durations and at optimal power can be used to increase both the early and 28-day strengths of mortar, cement paste, and concrete [2,12]. Figures 2.21 and 2.22 show the 3-, 7-, and 28-day compressive strengths of the mortar specimens cured with microwaves at two different power levels [2]. The optimal duration of curing to achieve both relatively higher early strength and 28-day strength can be identified as about 30 minutes for heating at 150 W and 15 minutes for heating at a microwave power of 300 W, resulting in about a 39%–47% increase in the 3-day compressive strength of mortar.

Microwave curing may also improve the strength of the concrete at an earlier age than those shown in Figures 2.21 and 2.22 [14]. Figure 2.23 shows that a typical mortar with a w/c ratio of 0.5 can achieve compressive strengths as high as 15 MPa in only 4.5 hours when microwave cured at an optimal microwave power (400 W in this case). The increased compressive strength development rate of the microwave-cured cement paste/concrete at early ages is mainly attributed to the increased rate of hydration of the microwave-cured cement. Figure 2.24 shows that microwave curing accelerates the hydration process during the first 24 hours [12]. The induction period of the specimens treated with microwaves is shorter than for conventionally cured cement paste, and the temperature in the second exotherm varies in proportion to the amount of energy absorbed by mortar. As seen in Figure 2.24, there is virtually no induction period in the mortar cured at the highest microwave power.

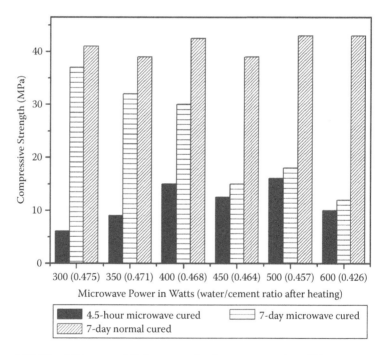

Figure 2.23 The 4.5-hour and 7-day strength of mortar cured at different microwave powers. All specimens were made using type III Portland cement. (From Leung, C.K.Y. and Pheeraphan, T., Very high early strength of microwave cured concrete. *Cement and Concrete Research*, 1995, **25**(1):136–146. With permission.)

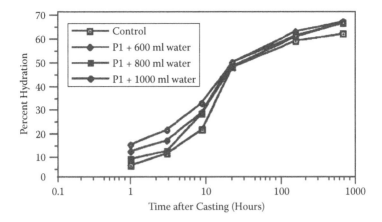

Figure 2.24 Percentage hydration versus time for mortar specimens microwave cured at different powers. Because of the lack of power metres, water loads were used to measure the microwave power delivered at each power setting. The relative value of the power levels can be considered as roughly proportional to the amount of water. (From Hutchison, R.G., Chang, J.T., et al., Thermal acceleration of Portland-cement mortars with microwave-energy, *Cement and Concrete Research*, 1991, **21**(5):795–799. With permisson.)

The increased hydration rate in microwave-cured mortar has also been confirmed by x-ray diffraction (XRD). XRD characterisation has indicated that the characteristic peaks of unhydrated C_3S and C_2S in microwave-cured mortar are lower than conventionally cured mortar specimens. The lower amounts of unhydrated C_3S and C_2S in the microwave-cured mortar corroborate observation of increased hydration rate caused by microwave curing. The collapse of capillary pores and densification of concrete because of the removal of water from the fresh mix are also considered to contribute to the increased early strength development rate achievable through microwave curing [2]. It has been shown that microwave curing can considerably reduce the permeability of concrete/mortar, leading to a denser microstructure in the microwave-cured mortar vis-à-vis the standard cured mortar, which can result in a stronger and more durable concrete/mortar (Figure 2.25) [2]. The densification of concrete mortar/concrete during microwave curing may be attributed to the plastic shrinkage and the resulting reduction in the porosity of mortar through removal of some of the free water at early ages.

As discussed, based on the results of a series of basic experimental studies, early investigations on microwave curing of concrete concluded that microwave heating generally increases both the early and long-term strengths of mortar/concrete. However, further studies indicated that the strength development trend of microwave-cured cementitious materials is

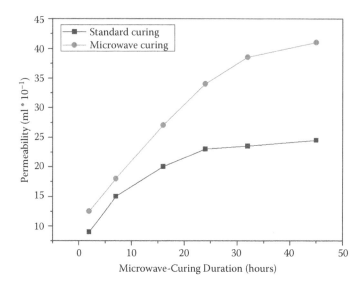

Figure 2.25 The variation in water permeability of mortar with time. A comparison between microwave-cured and standard-cured mortar. (Reprinted, with permission, from Wu, et al., Microwave curing technique in concrete manufacture. *Cement and Concrete Research*, 1987, **17**(2):205–210.)

not as easily predictable as initially supposed and changes with various process-specific parameters and mix characteristics. These include microwave power, type of cement used, and w/c ratio of the mix. Among these parameters, the effects of the w/c ratio and microwave power are significant and are discussed in detail in the subsequent sections. The type of cement, however, could also affect the strength development rate of the microwave-cured concrete. In 1995, Leung and Pheeraphan showed that, unlike the concrete and mortar made with type I cement, microwave-cured concrete/mortar specimens made with type III cement show a generally lower 7-day compressive strength than standard cured concrete/mortar (Figure 2.23). This is probably because of the higher heat generation of type III cement, which makes the specimens more sensitive to heating by microwaves [14].

In general, compared to mortar, microwave-cured concrete (with the same w/c ratio and mix characteristics) shows more improvements in the strength development rate. This is attributed to the expansion of the mortar during microwave heating and the formation of air bubbles because of localised boiling, which leads to the formation of microcracks and pores in the mortar. In concrete, the presence of aggregates restrains the expansion of the mortar, leading to the formation of lesser microcracks and pores in the concrete than in the case of mortar specimens [14].

2.5.2.1 Effects of water-to-cement ratio

The compressive strength development rate of concrete/cement paste/mortar subjected to microwave curing varies with the w/c ratio of the mix. Figure 2.26 shows that microwave curing of cement paste specimens with different w/c ratios at a constant microwave power can lead to very different compressive strength development profiles. This is because the water content of the concrete affects both the microwave heating rate and the availability of sufficient free water for hydration to occur. The very early strength development rates (up to 3 days) of microwave-cured cement paste/concrete may increase with a decrease in the w/c ratio. Figure 2.26 shows that a cement paste with a w/c ratio of 0.3 could achieve about 62.4% of its 28-day strength in 1 day, whereas this is about 54.2% for a cement paste with a w/c ratio of 0.5. As shown, the compressive strength of microwave-cured cement paste increases rapidly until 7 days and slows thereafter until a plateau is reached. The higher early strength development rate of cementitious mixes with lower w/c ratios (which usually contain higher cement contents) may be attributed to their higher dielectric loss. Rattanadecho et al. (2008) showed that cement paste specimens with a lower w/c ratio

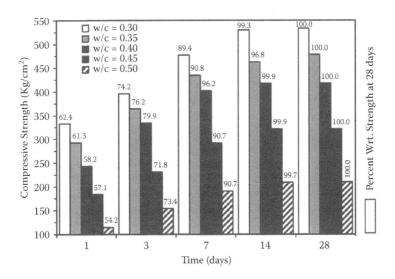

Figure 2.26 Compressive strength of cement paste specimens with different w/c ratios after microwave curing at 800 W for 15 minutes. (From Rattanadecho, P., Suwannapum, N., et al., Development of compressive strength of cement paste under accelerated curing by using a continuous microwave thermal processor. *Materials Science and Engineering*, 2008, **427**:299–307. With permission.)

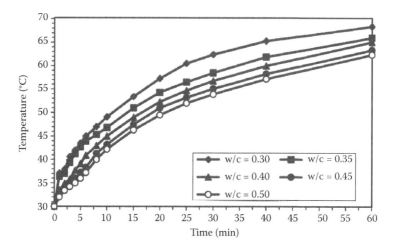

Figure 2.27 Variation in temperature of microwave-cured cement paste with w/c ratio at a constant microwave power. (From Rattanadecho, P., Suwannapum, N., et al., Development of compressive strength of cement paste under acceler-ated curing by using a continuous microwave thermal processor. *Materials Science and Engineering*, 2008, **427**:299–307. With permission.)

(or higher cement content) had greater microwave absorption and therefore were heated faster compared to mixes with higher w/c ratios (Figure 2.27) [21]. In addition, samples with a lower w/c ratio showed higher heat gen-eration from the hydration reaction, indicating that the additional heat from microwaves changes the kinetics of hydration in accordance with Arrhenius's law [22].

Besides the early strength development rate, the w/c ratio has also been reported to affect the long-term strength development rate of concrete. The reduction in the long-term strength of concrete/cement paste cured with microwaves is typically less at higher w/c ratios (Figure 2.28). This is attrib-uted to the availability of more free water in the microwave-cured concretes with higher w/c ratios, ensuring further long-term hydration of the cement.

2.5.2.2 Effects of microwave power

Microwave curing of concrete is generally performed at relatively low ISM microwave powers (<1000 W). This is partly because higher powers result in the expansion of specimens and overflowing of the slurry caused by the very rapid evaporation of water. The optimal microwave power for curing of a particular concrete mix, however, varies considerably with the char-acteristics of the mix, including w/c ratio, cement content, cement type, and so on. All these characteristics have overlapping effects that should be considered when selecting the appropriate microwave power for curing.

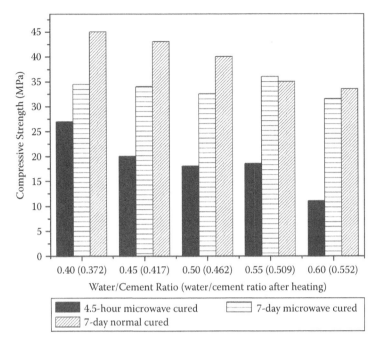

Figure 2.28 Effects of w/c ratio on compressive strength of concrete at 400 W. (From Leung, C.K.Y. and Pheeraphan, T., Very high early strength of microwave cured concrete. *Cement and Concrete Research*, 1995, **25**(1):136–146. With permission.)

Selection of microwave power is an important task in designing an optimal microwave-curing process. Figure 2.23 shows the early and long-term strengths of the mortar specimens with a w/c ratio of 0.5 cured using microwaves at different microwave powers. All samples were subjected to a constant total forward energy of 1.08 MJ by adjusting the heating duration for different microwave powers. This means that a higher microwave power was accompanied by a shorter heating duration to represent microwave curing with a higher heating rate. Type III cement was used to further accelerate the initial strength gain of the mortar. As shown, the long-term strength of mortar is more sensitive to rapid heating at an early age than the short-term strength. This is similar to the observations for the steam-cured concrete and suggests that the maximum microwave power level should be carefully selected to minimise any negative effects on the long-term strength of mortar/concrete. Figure 2.23 also shows that an increase in microwave power beyond 400 W results in a significant decrease in the 7-day strength of mortar while leading to few further improvements in early strength. For every concrete/mortar mix, there is a microwave power at which an acceptable compromise between the increase in early strength and decrease

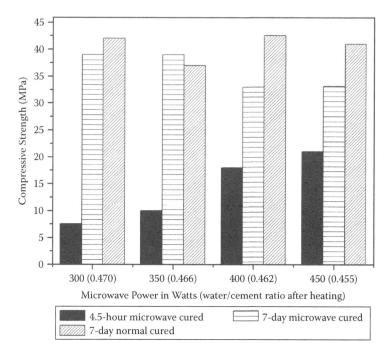

Figure 2.29 Effects of microwave power on compressive strength of concrete at 400 W. (From Leung, C.K.Y. and Pheeraphan, T., Very high early strength of microwave cured concrete. *Cement and Concrete Research*, 1995, **25**(1):136–146. With permission.)

in the long-term strength can be achieved. Figure 2.23 shows that, for the mortar samples tested, a microwave power of 400 W may provide the best compromise considering both early and later age strengths.

The effect of microwave power on the strength development of microwave-cured concrete is slightly different compared to the microwave-cured mortar (Figure 2.29). At lower powers, the 7-day strength of concrete shows relatively lesser sensitivity to microwave curing, and the microwave-cured specimens show relatively similar 7-day strengths as the standard-cured concrete specimens. Comparing Figures 2.23 and 2.29 shows that microwave-cured concrete specimens show less decrease in the 7-day compressive strength at 400 W and 450 W than the corresponding microwave-cured mortar specimens.

The majority of the experiments on microwave curing of cement paste, mortar, and concrete reported in the available literature have been performed by applying a fixed microwave power throughout the curing process. However, it has been debated that the use of constant power may lead to unnecessary overheating during the curing process. This is because of the variation in the dielectric properties of concrete/mortar as the concrete/mortar dries. Leung

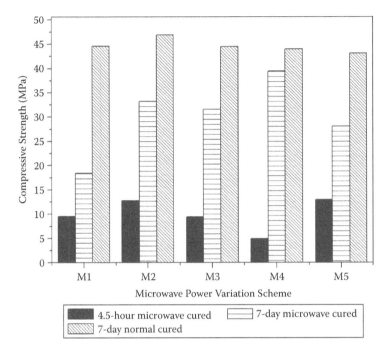

Figure 2.30 Compressive strength of microwave-cured mortar specimens subjected to various microwave power variation regimes. (From Leung, C.K.Y., and Pheeraphan, T., Determination of optimal process for microwave curing of concrete. *Cement and Concrete Research*, 1997, **27**(3):463–472. With permisson.)

and Pheeraphan recommended that a temperature-feedback control system be used to avoid unnecessary overheating by limiting the overall temperature reached in the specimens to below 80°C [15]. Figure 2.30 compares the compressive strength development of the mortar specimens that were microwave cured at five different microwave power variation regimes [15]. The first batch of mortar specimens (M1) were subjected to constant power at 412 W for 45 minutes. The second batch of mortar specimens (M2) were microwave cured for 90 minutes by limiting the temperature reached to below 60°C using a temperature-feedback system that regulates the power (maximum power of 600 W). The third batch (M3) was microwave heated up to 50°C in the first 20 minutes and up to 8°C in the next 25 minutes. The fourth batch (M4) was microwave cured for 45 minutes by limiting the temperature reached to 80°C (maximum power of 1200 W). The fifth batch (M5) of mortar specimens was cured at a constant power of 800 W for the first 20 minutes, followed by heating below the temperature limit of 80°C using the temperature-feedback system for another 25 minutes. The resulting average mortar temperatures reached throughout these curing processes

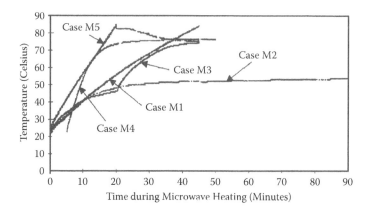

Figure 2.31 Temperature histories for various mortar specimens during microwave curing. (From Leung, C.K.Y., and Pheeraphan, T., Determination of optimal process for microwave curing of concrete. *Cement and Concrete Research*, 1997, **27**(3):463–472. With permission.)

are compared in Figure 2.31. As shown in Figure 2.30, the outcome of the microwave-curing process could vary significantly depending on the curing regime used. The M1 and M4 regimes seemed to provide the best combinations of early and later age strengths. When two or more regimes result in relatively similar outcomes in terms of compressive strength development, other criteria, such as energy consumption, can be used to select the optimal-curing process. For the cases discussed, the energy consumption of regime M1 and M4 can be estimated as 0.309 kWh and 0.302 kWh, respectively, using the area of power-versus-time curves for the two regimes [15].

Similar to mortar, microwave curing of concrete is also sensitive to the microwave power regime used. Figure 2.32 shows the variation in the compressive strength of concrete specimens (w/c = 0.4) subjected to four different microwave power regimes. In the first regime (C1), concrete specimens were cured at a constant microwave power of 412 W for 45 minutes. The second batch of concrete (C2) was microwave cured for 45 minutes by limiting the temperature reached to below 80°C using the temperature-feedback system (maximum power of 750 W). The third batch of concrete (C3) was cured for 45 minutes by limiting the maximum temperature to 85°C (maximum power of 1200 W). The fourth batch of concrete (C4) was microwave cured for 45 minutes at a controlled temperature determined using the power-versus-time curve for C2. In this regime, a reversed history with power at time *t* for C4 equal to power at time (45 minutes – *t*) for case C2 was applied. With this method, C4 concrete specimens receive the same amount of energy as C2 but less energy during the earlier stage of microwave curing and more at the later stage. As shown in Figure 2.32, both

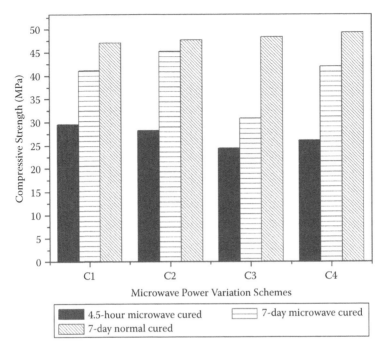

Figure 2.32 Compressive strength of microwave-cured concrete specimens subjected to various microwave power variation regimes. (From Leung, C.K.Y., and Pheeraphan, T., Determination of optimal process for microwave curing of concrete. *Cement and Concrete Research*, 1997, **27**(3):463–472. With permission.)

early and late strengths of regime C4 are lower than regime C2. While both regimes (C2 and C4) reach relatively similar temperatures after curing, the more rapid curing at early ages in C2 seems to be closer to the optimal process. Compared to C2, C1 has higher early age strength but lower later age strength. Moreover, the high heating rate of regime C3 results in considerable reduction in the later age strength of the concrete. By summing the percentage of the reference strength achieved at 4.5 hours and 7 days, the overall strength development pattern of C2 seems to be slightly better than C1. In addition, from power-versus-time curves, it can be determined that the total power consumption of C2 is about only 0.241 kWh, which is considerably lower than that of C1 (0.309 kWh). This indicates that the feedback-temperature control system can be used to achieve relatively similar early and later age strengths compared to the constant power regime with about 20% lower energy consumption. Therefore, it can be concluded that although the use of the feedback control system to regulate power during the curing regime rather than microwave curing at constant power results in only minor variations in the strength development rate of concrete, this is useful as an energy-saving method.

2.5.2.3 Effects of initial delay

It is common to delay the start of the microwave-curing process and allow the concrete/mortar to cure normally for a short period of time known as the delay period. Delaying the start of the curing process has also been recommended in most conventional accelerated curing methods, such as steam curing, to improve the effects of the curing process. Figure 2.33 shows the variation in the strength development trend of microwave-cured mortar specimens (w/c = 0.5) with the duration of the delay period. As shown, delaying the start of microwave heating can have considerable effects on the early strength (4.5 hours) of the microwave-cured mortar/concrete. The optimal delay time for the specimens as observed in Figure 2.33 can be easily determined to be about 30 minutes [14]. Similarly, it is important to identify the effects of the delay in the start of microwave curing on the

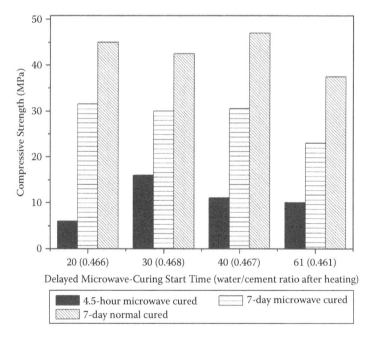

Figure 2.33 Effect of delayed microwave curing on the compressive strength of microwave-cured mortar specimens. (From Leung, C.K.Y., and Pheeraphan, T., Determination of optimal process for microwave curing of concrete. *Cement and Concrete Research*, 1997, **27**(3):463–472. With permission.)

strength development of various mixes with different compositions and w/c ratios prior to microwave curing of actual precast concrete components.

2.5.2.4 Effect of microwave-curing duration

Identifying the optimal duration of microwave curing to achieve the desired strength development rate is crucial to prevent overheating. Excessive microwave curing can negatively affect the long-term properties of the concrete and result in unnecessarily high energy consumption and associated emissions. Figure 2.34 compares the compressive strength development curve of the concrete specimens microwave-cured for 15 minutes and 30 minutes at similar constant microwave powers. As shown, not only does excessive microwave curing not result in any improvement in early concrete strength development rate, but it may also negatively affect the long-term strength of the microwave-cured samples. This is probably because an excessively long microwave-curing duration leads to a lack of free water available for hydration and an increase in capillary voids. The reduction in the w/c ratio of mortar specimens subjected to microwave heating as reported by Wu et al. (1987) is shown in Table 2.4.

Based on the results summarised in this table and Figures 2.21 and 2.22, one may conclude that the compressive strength of mortar (with an initial w/c ratio of 0.5) decreases when its final w/c ratio decreases to less than 0.40.

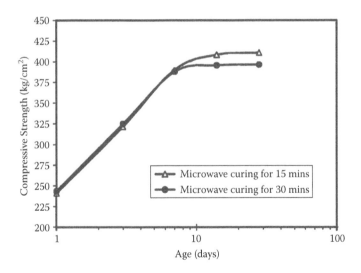

Figure 2.34 Effect of extended microwave curing on the compressive strength of cement paste. (From Rattanadecho, P., Suwannapum, N., et al., Development of compressive strength of cement paste under accelerated curing by using a continuous microwave thermal processor. *Materials Science and Engineering*, 2008, **427**:299–307. With permission.)

Table 2.4 w/c Ratio (after microwave curing) versus the duration
          of microwave curing

Microwave power (watts)	Curing duration					
	0 min	15 min	30 min	60 min	90 min	120 min
150	0.44		0.416	0.342	0.290	0.222
300	0.44	0.404	0.351	0.215	0.110	

This is consistent with the theory of hydration, which indicates that a w/c ratio equal to or greater than 0.42 is required to ensure the availability of sufficient free water to achieve complete hydration of the cement. Powers et al. (1958) showed that, at w/c ratios of about 0.38, all the water will be in the form of combined water and gel water, and no capillary water will be available if the cement is fully hydrated [23,24]. This means that at w/c ratios below 0.38, achieving full hydration is difficult. Therefore, fine-tuning the microwave-curing process to prevent excessive heating is critical to maintain the final w/c ratio of the mix at levels sufficient to have free water available for hydration.

2.6 SUMMARY

Accelerated curing at elevated temperatures is an effective method to increase the early strength development rate of concrete, which is highly appealing to the precast concrete industry. However, conventional elevated-temperature accelerated curing methods, including low-pressure steam curing, high-pressure steam curing, and conduction/convection heating-based curing methods, have a number of major drawbacks that limit their widespread application in the precast concrete industry. Conventional elevated-temperature curing methods usually result in some nonuniformity in the degree of hydration of cement throughout the concrete element because of the inherent thermal insulating properties of concrete and the difficulty of conventional heating sources in achieving uniform heating. Nonuniform heating of concrete may result in the development of differential thermal stresses that negatively affect the long-term properties of the concrete by introducing new nano- and microcracks and accelerating the growth rate of the existing cracks present in the concrete. In addition, another important disadvantage of conventional accelerated curing methods is their relatively long processing durations. Even the shortest conventional accelerated curing processes, such as steam curing, require about 10 hours to complete before demoulding is possible.

Microwave curing is believed to have great potential to revolutionise the curing of precast concrete. Microwave heating at lower ISM frequencies can be used to heat the concrete uniformly and significantly reduce the time required for curing. Unlike steam-cured concrete, microwave-cured concrete can gain considerable strengths in just a few hours without compromising its long-term properties.

This chapter reviewed the recent achievements in development and application of microwave curing. The working principles of microwave curing were discussed to enhance understanding of the phenomenon leading to the increased strength development rate observed in the microwave-cured concrete. It was shown that an optimal duration of microwave curing at an optimal microwave power can result in considerable improvements in the early strength development rate of concrete without having a negative effect on its long-term properties. It was shown that the outcome of the microwave-curing process was affected considerably by various process-specific parameters and concrete characteristics, including microwave power, microwave heating duration, and the w/c ratio of the concrete. These parameters have overlapping effects that eventually have a bearing on the properties of the microwave-cured concrete produced and therefore should be considered carefully when designing the microwave-curing processing regime for a particular concrete mix.

REFERENCES

1. Mindess, S., Young, J.F., and Darwin, D., *Concrete*. 2nd edition, Upper Saddle River, NJ: Pearson Education, 2003.
2. Wu, X.Q., Dong, J.G., and Tang, M.S., Microwave curing technique in concrete manufacture. *Cement and Concrete Research*, 1987, 17(2):205–210.
3. Gerwick, B.C.J., *Construction of Prestressed Concrete Structures*. 2nd edition. New York: Wiley, 1993, 19–23, 91–94.
4. Heritage, I., Khalalf, F.M., and Wilson, J.G., Thermal acceleration of Portland cement concretes using direct electronic curing. *ACI Materials Journal*, 2000, 97(1):37–40.
5. Tepponen, P. and Eriksson, B.-E., Damages in concrete railway sleepers in Finland. *Nordic Concrete Research*, 1987, 6:199–209.
6. French, C., Mokhtarzadeh, A., et al., High-strength concrete applications to prestressed bridge girders. *Construction and Building Materials*, 1998, 12(2–3):105–113.
7. Detwiler, R.J. and Mehta, P.K., Chemical and physical effects of silica fume on the mechanical-behavior of concrete. *ACI Materials Journal*, 1989, 86(6):609–614.
8. Detwiler, R.J., Fapohunda, C.A., and Natale, J., Use of supplementary cementing materials to increase the resistance to chloride-ion penetration of concretes cured at elevated-temperatures. *ACI Materials Journal*, 1994, 91(1):63–66.
9. Levitt, M., *Precast Concrete, Materials, Manufacture, Properties and Usage*. Englewood, NJ: Applied Science, 1982, 33–38, 53–73.

10. Kosmatka, S.H. and Panarese, W.C., *Design and Control of Concrete Mixtures*. Skokie, IL: Portland Cement Association, 1988, 66–67.
11. Hester, W.T., High-range water-reducing admixtures in precast applications. *PCI Journal*, 1978, **July–August**:68–85.
12. Hutchison, R.G., Chang, J.T., et al., Thermal acceleration of Portland-cement mortars with microwave-energy. *Cement and Concrete Research*, 1991, **21**(5):795–799.
13. Bella, G.D., S. Lai, and M. Pinna, Microwaves for the peraccelerated curing of concretes. *Betonwerk und Fertigteil-Technik/Concrete Precasting Plant and Technology*, 1994, **60**(12):87–93.
14. Leung, C.K.Y. and Pheeraphan, T., Very high early strength of microwave cured concrete. *Cement and Concrete Research*, 1995, **25**(1):136–146.
15. Leung, C.K.Y. and Pheeraphan, T., Determination of optimal process for microwave curing of concrete. *Cement and Concrete Research*, 1997, **27**(3):163 172.
16. Mak, S.L., Microwave accelerated processing for precast concrete production. In *Proceedings of the 4th CANMET/ACI International Conference on Durability of Concrete*, Sydney, Australia, August 17–22, 1997, pp. 709–720.
17. Mak, S.L., Accelerated heating of concrete with microwave curing. In *Proceedings of the 4th CANMET/ACI/JCI International Conference on Recent Advances in Concrete Technology*, Tokushima, Japan, June 7–11, 1998, pp. 531–542.
18. Sohn, D.G. and Johnson, D.L., Microwave curing effects on the 28-day strength of cementitious materials. *Cement and Concrete Research*, 1999, **29**(2):241–247.
19. Mak, S.L., Banks, R.W., et al., Advances in microwave curing of concrete. In *Fourth World Congress on Microwave and Radio Frequency Applications*, Sydney, Australia, September 22–26, 2002. Available at HYPERLINK "https://publications.csiro.au/rpr/pub?list=BRO&pid=procite:76309505-ab72-4edb-a473-dacba4229de8" https://publications.csiro.au/rpr/pub?list=BRO&pid=procite:76309505-ab72-4edb-a473-dacba4229de8
20. Lee, M.G., Preliminary study for strength and freeze-thaw durability of microwave- and steam-cured concrete. *Journal of Materials in Civil Engineering*, 2007, **19**(11):972–976.
21. Rattanadecho, P., Suwannapum, N., et al., Development of compressive strength of cement paste under accelerated curing by using a continuous microwave thermal processor. *Materials Science and Engineering*, 2008. **427**:299–307.
22. Makul, N., Agrawal, D.K., and Chatveera, B., Microstructures and mechanical properties of Portland cement at an early age when subjected to microwave accelerated-curing. *Journal of Ceramic Processing Research*, 2011, **12**(1):62–69.
23. Powers, T.C., Structure and physical properties of hardened portland cement paste. *Journal of the American Ceramic Society*, 1958, **41**(1):1–6.
24. Powers, T.C., Absorption of water by Portland cement paste during the hardening process. *Industrial and Engineering Chemistry*, 1935, **27**:790–794.

Chapter 3

Microwave-assisted selective demolition of concrete

3.1 INTRODUCTION

Partial removal of concrete and drilling into concrete are inevitable in many retrofitting, repair, restoration, and deconstruction projects. However, because of the considerably high toughness of concrete, it is usually difficult to demolish or drill into a particular section of concrete components without causing damage to the surrounding concrete. Selective demolition of concrete using conventional tools, including chisels and hammers, pneumatic breakers, and hydraulic breakers, poses considerable health hazards to the equipment operators, building occupants, and other affected individuals because of the relatively high amount of noise, dust, and vibration generated during the demolition process. Heightened awareness about the social element of sustainability in construction has highlighted the necessity of eliminating possible causes of health hazards in construction sites as well as reducing the noise and dust pollution caused by construction activities. The inevitable tightening of safety and health standards together with the constant push for efficiency in the construction industry have motivated a number of researchers to seek alternative selective demolition techniques for concrete. The recent advances in technology have opened up a variety of technological options to develop new selective demolition techniques for concrete with reduced noise and dust generation, through avoiding mechanical collisions and improved removal rates.

This chapter looks into applications of microwave heating in the selective demolition and drilling of concrete. The chapter starts by reviewing a number of commonly encountered situations for which the use of more efficient selective demolition tools (e.g., microwave removal and drilling) can make a difference. It then continues by reviewing the designs and working principles of a number of potential selective demolition tools proposed in the available literature and describes the typical components of such tools. The focus is placed on microwave-assisted selective demolition and microwave-assisted drilling methods as alternatives to the existing selective demolition and drilling methods for concrete. The results of a numerical study considering representative concrete specimens subjected to microwave-assisted selective demolition are presented to facilitate understanding of the underlying phenomena leading to the demolition of concrete using the proposed microwave-assisted methods.

3.2 APPLICATIONS OF SELECTIVE CONCRETE REMOVAL TECHNIQUES

Many civil engineering projects require the partial removal of concrete. This can come about when the concrete surface has lost its strength and functionality because of exposure to hostile physical or chemical environments. In the following, a few examples of such cases are reviewed.

3.2.1 Removal of radioactive or chemically contaminated concrete surface

One of the most important applications of highly selective concrete removal techniques is in removal of the contaminated surface of concrete that has been used in nuclear power plants, nuclear waste-processing plants, or storage and processing facilities for other types of hazardous materials. For instance, when concrete is used in shielding or as structural elements in nuclear power plants (Figure 3.1), as a consequence of long-term usage, various radionuclides diffuse into the concrete, contaminating the surface

Figure 3.1 Concrete is a common material in nuclear power plants and nuclear waste-processing plants.

layer. The thickness of the contaminated layer depends on the concrete diffusivity and exposure duration and is usually between 1 and 10 mm [1]. This thin contaminated layer should be selectively removed and segregated from uncontaminated concrete during the decommissioning or retrofitting of nuclear power plants and disposed under special precautions required for the disposal of hazardous wastes. Disposal of radioactive-contaminated waste is a sophisticated and costly process as the waste poses risks to humans and wildlife.

In the case of concrete subjected to radionuclides, the contamination is usually limited to a thin surface layer. If a selective demolition method for removal of the contaminated surface layer of concrete is not available, difficulties in separating the contaminated from uncontaminated parts of the concrete are evident. If not removed fully, disposal of a much more concrete waste would ensue, adding to much higher decommissioning costs. Aside from decommissioning costs, disposal of a sizable volume of hazardous waste is more challenging and is expected to pose more severe environmental issues (e.g., isolation of disposal sites and the impact on wildlife habitats and the natural ecosystem). The lack of selective concrete demolition techniques to remove the contaminated surface layer of concrete without affecting the integrity of the remaining concrete also adds to the economic and environmental impact of decommissioning and retrofitting by requiring additional effort (energy use and associated emissions) for complete demolition and reconstruction of the affected elements.

As seen in this example, the availability of an efficient technique to demolish selectively a specified thickness of the concrete surface layer without affecting the integrity of the remaining concrete can lead to significant economic and environmental benefits.

3.2.2 Removal of chloride-contaminated and deteriorated sections of concrete

Deterioration caused by chloride is a serious problem, especially in concrete structures located in regions with fluctuating water levels and in offshore structures where such marine environments provide a continuous supply of salts, water, and oxygen. Similarly, concrete in contact with groundwater that is rich in salts may also be prone to chloride deterioration. In addition, to keep roads safe for travel during winter, sodium and calcium chloride are commonly used as efficient and cheap deicing agents. The salts used for deicing dissolve in the snow and slush, eventually ending up as runoff water that penetrates into the concrete, creating a hostile environment for the reinforcing bars.

The penetration of salts into concrete may damage the surface through two phenomena. First, chloride ions can destroy the passive oxide film on steel reinforcing bars, even at high alkalinities, and thereby accelerate steel

Figure 3.2 Concrete deteriorated because of chloride contamination. (Courtesy of James Bushman, Bushman & Associates, Inc., 6395 Kennard Road, PO Box 425, Medina, OH 44256 USA.)

corrosion. The formation of rust caused by the corrosion is an expansive reaction and can result in cracking and spalling of the concrete cover of the corroding steel (Figure 3.2). Second, the salt may deteriorate the concrete surface through crystallisation. Crystallisation of salts may damage the concrete through the development of crystal growth pressures. As a result of evaporation, the salts are concentrated and deposited in the concrete pores and may continue to the point at which they cause cracking.

To rehabilitate structures exposed to chloride attack, some amount of concrete removal and replacement is usually necessary. Considering the amount of the surface area involved, removal of surface concrete using conventional concrete removal techniques is time consuming, energy intensive and laborious, and usually results in damage to the underlying concrete and may even compromise the stability of the structure as a whole. An efficient method to remove the surface concrete without causing damage to the underlying concrete is highly appealing if available.

3.3 STATE-OF-THE-ART SELECTIVE CONCRETE DEMOLITION TECHNIQUES

To better understand the need for developing more efficient selective demolition tools for concrete, it is important to gain some level of understanding about the advantages and disadvantages of available state-of-the-art selective demolition technologies. A number of viable technologies for the selective demolition of concrete are reviewed in this section. It is important to note that only a few of the following techniques have actually been deployed in practice, and the majority of the technologies reviewed require further investigation to verify efficiency, effectiveness, as well as the actual economic and environmental impacts when used as selective concrete demolition tools. The selective demolition technologies available vary widely in their effectiveness and processing rates. Except for demolition using chemical agents (expansive agents), other technologies discussed in the following material rely mainly on the mechanical removal of concrete using cutting and sawing equipment. As a result, most of the following methods share the main disadvantages associated with mechanical-based methods, including the associated health hazards and safety issues.

3.3.1 Pneumatic concrete breakers

Pneumatic breakers (Figure 3.3) are probably the most common type of breakers for selective demolition in construction. In pneumatic breakers, a pneumatic actuator is used to convert the compressed air's energy into mechanical motion. The advantages of pneumatic breakers compared to other available breakers include being relatively lightweight and inexpensive. They are also considered long lasting and fairly easy to repair.

However, disadvantages include the requirement for fairly large air compressor power sources, noise and dust hazards, and inoperability at low temperatures. In addition, such air tools usually require a fair amount of maintenance, including lubrication and the need to avoid moisture buildup in lines. The use of breakers also poses health hazards caused by vibration. Vibration is recognised to cause various health hazards, including blanching fingers under cold provocation, the so-called white fingers.

Figure 3.3 Handheld pneumatic breaker.

Figure 3.4 Handheld hydraulic breaker.

3.3.2 Hydraulic breaker

Handheld hydraulic breakers are also commonly used for the selective removal of concrete. Such breakers are powered by an auxiliary hydraulic system that delivers pressurised hydraulic fluids from a hydraulic pump. A handheld hydraulic breaker is shown in Figure 3.4. In fact, both pneumatic and hydraulic breakers are similar in the sense that they both use fluid power. Pneumatic breakers use an easily compressible gas such as air; hydraulic breakers use relatively incompressible liquid media such as oils. Compared to pneumatic breakers, the main advantage of hydraulic breakers is the fact that, unlike the gases used in pneumatic breakers, the liquid used in hydraulic breakers do not absorb any of the supplied energy. Hydraulic breakers are usually capable of providing much higher breaking forces because of fluid incompressibility, which also minimises spring action.

3.3.3 Expansive agents

Expansive demolition agents are an alternative to conventional mechanical demolition tools such as hydraulic and pneumatic breakers and explosives. As shown in Figure 3.5, an array of holes is drilled in selected portions of the concrete element in order to break up the concrete using expansive agents. The holes are then filled with a slurry mixture of the expansive demolition agent and water. The slurry expands gradually and results in relatively high expansive stresses within the holes, which crack the concrete.

The main advantages of demolition using expansive agents are the reduced noise and dust pollution as they are silent and do not produce vibration.

Figure 3.5 Selective demolition of concrete using expansive chemical agents.

Furthermore, this method is relatively safer than other conventional demolition methods and may be more economical in certain situations.

The main disadvantage of this method is the relatively long time required for expansion to occur. Furthermore, the drilling of holes itself has to be done using conventional methods, which is time consuming with the same disadvantages associated with conventional demolition techniques.

3.3.4 Hydrodemolition

Hydrodemolition is another common method for the selective demolition of concrete. This method uses high-pressure water jets to delaminate the selected surface area of the concrete component. Hydrodemolition is particularly suitable for use in concrete repair applications when a particular area of surface concrete is to be replaced with a new layer of concrete. This is because, after demolition, the surface of the remaining concrete is usually left roughened. The rough surface of the concrete substrate enhances the bond and the composite action between the substrate concrete and the new concrete.

Hydrodemolition is considered to be a relatively fast and efficient method for the selective demolition of concrete. The use of water in this method can provide a healthier environment by suppressing the amount of dust generated. Hydrodemolition is also particularly effective in separating concrete from the reinforcing bars and other embedded steel elements in concrete.

However, hydrodemolition also has considerable drawbacks and limitations. One of the main drawbacks of hydrodemolition is related to the huge amount of wastewater produced and the need for wastewater disposal. The wastewater produced during hydrodemolition contains fine cement particles with a pH ranging from 11 to 13 and thus requires special measures prior

Figure 3.6 Concrete removal using hydrodemolition. (Reprinted, with permission, from ACIRAP-14reportpublishedbyACICommitteeE706:AmericanConcreteInstitute, Field Guide to Concrete Repair Application Procedures: Concrete Removal Using Hydrodemolition. Farmington Hills, MI: ACI, 2010.)

to disposal through local sanitary systems. Furthermore, hydrodemolition is not suitable for removing concrete in posttensioned structural elements as water may enter the posttensioned tendon sheathing and cause long-term durability problems. Apart from these issues, the biggest disadvantage associated with hydrodemolition is the safety issues arising from the use of this method. The water jet is dangerous and can cause serious injury and even death. Water jets can propel debris at high velocity and may result in injury or death as well as damage to property. The hydrodemolition tool should only be operated by a highly experienced operator (Figure 3.6).

3.3.5 Abrasive jetting

Abrasive jetting (Figure 3.7) uses a high-speed stream of particles carried in an air or gas jet to remove a thin layer (paint, concrete, rust) from the

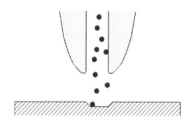

Figure 3.7 Abrasive jetting.

surface of the material [2]. The abrasive material is normally aluminium oxide or silicon carbides. Abrasive particles should have sharp edges with diameters of about 10–50 μm. This process is most effective on flat surfaces and can be used on "hard-to-reach" areas such as ceilings. The main disadvantage of this process is the comparatively large amount of secondary waste produced [3]. In addition, the abrasive jetting removal method has been reported to be effective only when a thin layer of the surface is to be removed.

One variation of the abrasive jetting technique, known as high-pressure liquid nitrogen blasting, uses liquid nitrogen as the jet material [2]. In this method, the concomitant effect of the embrittlement caused by liquid nitrogen and the abrasive action of the grit is expected to result in slight improvements in removal efficiency.

Wet ice blasting, in which a compressed air jet is used to propel a mixture of water and ice crystals onto the surface to be removed, is another variation of abrasive jetting. [3].

3.3.6 Other methods

The most commonly used methods for the selective removal of concrete were reviewed in the previous sections. However, in addition to these methods, various other removal technologies are available that could have potential for the selective removal of concrete. These methods include sponge blasting, CO_2 blasting (dry ice blasting), electrohydraulic scabbing, shot blasting, soda blasting, and laser ablation. The surface removal rates of these methods range from 1 to 20 m^2/h, depending on the method and the type of the equipment used. Although these methods have been used in a variety of other applications, they have seldom been used in practice to remove concrete. The application of such methods for the selective removal of concrete requires further investigation.

3.4 MICROWAVE-ASSISTED DEMOLITION OF CONCRETE

Microwave heating recently has been considered as an efficient alternative to conventional methods in the selective demolition of concrete. As described in Chapter 1, the microwave penetration depth in the materials and the microwave absorption by materials vary with microwave frequency and the dielectric properties of the materials. As a result, microwaves can be used to heat the materials selectively through choice of an appropriate microwave frequency range for a particular penetration depth in a given material. Appropriate design of the microwave applicator will enable the user to focus the microwave energy onto the desired area. This unique selectivity renders microwaves as a promising candidate for developing selective demolition tools. Microwave-assisted selective demolition of concrete has been proposed recently as an efficient technique for the partial or complete removal of a selected portion of concrete components with minimal adverse effects on adjacent portions, making it considerably more desirable, especially for renovation and rehabilitation purposes [4–8]. Microwave-assisted selective demolition of concrete can reduce considerably the health and safety concerns associated with conventional selective demolition methods, including noise and dust-related hazards. This is because microwave-assisted selective demolition relies mainly on relatively quiet demolition of concrete through concentrating the energy in the demolition zone without the need for any mechanical engagement between the demolition tool and the concrete.

In Chapter 2, we observed that microwave curing of concrete mainly makes use of lower industrial, scientific, and medical (ISM) frequencies (<2.45 GHz) to ensure the uniform heating of concrete components, thereby avoiding the development of detrimental thermal stresses. Although potential thermal stresses caused by differential heating of the materials using microwaves of higher ISM frequencies are not desirable and should be prevented in the microwave curing of concrete, in selective demolition of concrete such stresses may be harnessed to cause delamination of the microwave-exposed region. With this in mind, higher microwave frequencies (>2.45 GHz) are usually more appropriate for microwave selective demolition methods. The higher the microwave frequency, the more localised the microwave power would be, resulting in higher selectivity and localisation of the thermal stresses in the microwave-exposed area of the concrete component. Such localised heating has been shown to result in differential thermal stresses and elevated pore water pressures that can be harnessed to remove the heated parts of the concrete.

The idea of using microwaves to develop selective demolition tools for concrete first appeared in 1968, in a book chapter written by A. Watson [9], who invented a device for cracking structures made of concrete, brick, or the like through microwave heating. They reported that the invented device uses microwave energy that passes along a waveguide to form a narrow beam of energy, causing it to emerge from the waveguide through an opening having a width substantially less than that of any part of the waveguide. Directing this beam at a small area on the surface of a structure for a sufficient amount of time causes such localised heating that it results in the cracking of the structure. The inventors recommended that the best performance may be achieved when the device uses frequencies higher than 0.8 GHz. It was reported that a temperature rise of at least 200°C at a depth of 7.5 cm and a temperature rise of at least 400°C at the surface of the structure may be reached before cracking occurs [10].

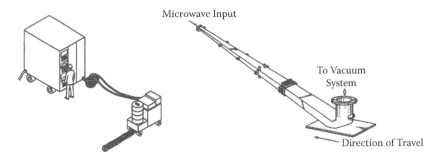

Figure 3.8 Mobile system for microwave removal of concrete surfaces invented by White et al. (1997) [11].

Subsequently, several research groups, including Japan's Atomic Energy Research Institute (JAERI), UK's Harwell Laboratory, and the Oak Ridge National Laboratory (ORNL) in the United States, tried to develop methods and the necessary equipment to use microwaves as a demolition tool to remove radioactive wastes from the surfaces of concrete. In 1997, White et al. invented a mobile system for the microwave-assisted removal of concrete surfaces [11]. The apparatus was reported to use the thermal stresses generated through microwave heating alone to etch the concrete surface to the desired depth [12,13]. The configuration of the primary microwave-assisted concrete etching device invented by White et al. is shown in Figure 3.8. The apparatus comprises a housing adapted to pass over a support surface. The housing includes a waveguide for directing microwave energy to the surface of concrete at an angle that maximises the absorption of microwave energy by the surface. The apparatus as reported is operated by a source of microwave energy matching the associated waveguide, wherein the microwave has a frequency of between about 10.6 GHz and about 24 GHz targeted to remove the uppermost layers from the surface. The apparatus also contains a debris containment assembly and vacuum assembly for convenient operation with the housing. As can be seen, White et al.'s invention operates at a considerably higher microwave frequency range (10.6–14 GHz) compared to the first-reported microwave-assisted demolition device invented by Watson (minimum recommended microwave frequency = 0.8 GHz). Such an increase in the proposed microwave frequency range to achieve better results stems from the significant technological development within the period of time between the two inventions with the availability of considerably cheaper high-frequency microwave sources. Furthermore, this could also be due to a better understanding of the behaviour of microwaves in concrete, appreciating the fact that higher microwave frequencies are more effective for selective demolition purposes.

Apart from concrete, microwave heating has also been proposed for use as a selective demolition tool for other hard dielectric materials. This includes applications of microwaves in the breakage of hard rocks in the mining industry. A considerable amount of research and development have been done in applying microwave radiation to rocks like granite, schist, pumice, slate, and sandstone. The results of such studies have shown that microwave heating can effectively soften the rock and eventually lead to surface delamination. However, it has also been reported that the microwave-assisted rock demolition tools that rely solely on microwave energy tend to be uneconomic when deployed in the mining industry. This is mainly because of the highly integrated nature of hard rocks, which requires an excessive amount of microwave energy to cause demolition. Therefore, various combinations of microwave heating and mechanical cutting have also been proposed, being more economic and more environmentally friendly approaches for the selective demolition of hard rocks.

In 1966, Puschner invented a microwave heating device for rock breakage; it is shown schematically in Figure 3.9 [14]. The objective of this invention was to provide an efficient tool to cleave large blocks of rock and ore for the purpose of subsequent comminution. The proposed tool consists of a microwave generator, a cooling device for the generator, and a hollow-guide radiator for directing the microwave energy onto the material to be heated. The microwave device is mounted on the boom of a crane to enable application of microwave energy through its hollow-guide radiator onto the material to be heated. The inventors have claimed that this movable apparatus has high flexibility, providing easy access to the hard-to-reach edges of rocks [14].

Another microwave-assisted apparatus for demolition of hard rocks was invented in 1991 by Lindroth et al. [15]. The apparatus was reported to be especially suitable for sequential fracturing and cutting of the subsurface volume of hard rocks in strata typical of a mining environment. The method involves subjecting the volume of rock to a beam of microwave energy to fracture the subsurface volume of the rock by differential expansion and then bringing the cutting edge of a piece of conventional mining machinery into contact with the fractured rock. A schematic presentation of this apparatus is shown in Figure 3.10. The inventors claimed that the combination of microwave-assisted and mechanical demolition tools used in this apparatus may lead to less wear of the equipment compared to when only microwave heating or only conventional mining methods are used [15].

In 1993, Lindroth et al. reported a series of experimental test results using the combined demolition tool. The experiments were conducted on two selected igneous rocks, namely, dresser basalt and St. Cloud grey granodiorite. During the entire test, the drilling parameters were held constant, with variables being microwave power and time of irradiation. The 401-kg drill thrust, rotation of 36 rpm, and 1-minute drilling time were

Figure 3.9 Microwave heating device for breakage of rocks invented by Puschner (1966) [14].

also held constant. Furthermore, a high-power microwave generator providing power of up to 25 kW at a frequency of 2.45 GHz was used. All the experiments were performed inside a closed copper screen room used for containing microwave energy. The results of the experiments showed that microwave heating can significantly increase the penetration depth rates of the mechanical tools in selected types of hard rock.

3.5 CONFIGURATION OF MICROWAVE-ASSISTED SELECTIVE DEMOLITION TOOLS

A number of available microwave-assisted selective demolition technologies (applicable to concrete demolition) were reviewed in the previous section. Microwave-assisted equipment will be discussed next.

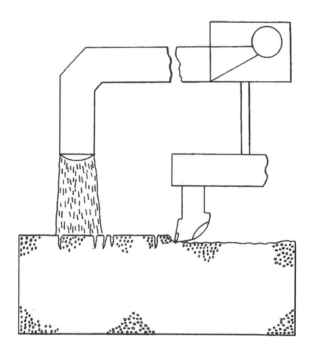

Figure 3.10 Microwave-assisted fracturing and cutting device for hard rocks invented by Lindroth et al. (1991) [15].

3.5.1 Microwave source and generator

The microwave source and generator are responsible for supplying the microwave energy required. The most commonly used source of microwave energy, primarily for reasons of efficiency, is the magnetron (Figure 3.11). Because of mass production, magnetrons at 2.45 GHz are particularly cheap; however, magnetrons for other frequencies are also available. Other microwave sources are available, such as the travelling wave tube (TWT), klystron, gyrotron, and solid-state devices. Each of these has characteristics that can be exploited. For instance, at higher frequencies (e.g., 10.6 and 18 GHz), the klystron and gyrotron are normally more efficient than magnetrons.

3.5.2 Microwave transmission line

The microwave transmission line is responsible for transmitting the microwave energy from the microwave source to the microwave applicator. This unit comprises a variety of waveguide components used to deliver the generated microwave power to the applicator while minimising the power

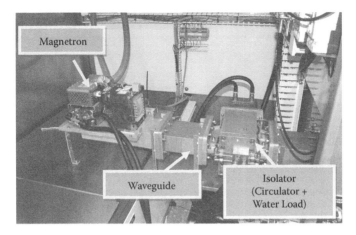

Figure 3.11 Magnetron, waveguide, and isolator in an industrial microwave system.

reflection. Waveguides are hollow metallic objects that confine the waves to the space within their walls and guide the waves from one point to another. A hollow metallic tube of either rectangular or circular cross section made of aluminium, copper, or brass of various sizes is generally used in practice. Because of the boundary conditions imposed by the waveguides, the waves that propagate inside a homogeneously filled waveguide are different from the waves propagating in free space (discussed in detail in Chapter 6). Other components of the microwave transmission line usually include the microwave isolator, microwave power-monitoring unit, and tuner. The isolator unit is an important component in the transmission lines to minimise the risk of damage to the generator unit caused by the reflected power. The isolator is usually a two-port device made of a ferrite material and magnets that does not permit the flow of power in the reverse direction (Figure 3.11). The monitoring unit is an optional component of the transmission line and is responsible for estimating the generated and reflected power by taking samples from the microwaves being transmitted in both directions. The tuner is responsible for maximising the power transfer from the source to the load (concrete) through a process known as impedance matching and is important in improving the efficiency of the microwave-assisted demolition tools. The role and configuration of each of the components used in the transmission line of typical microwave heating systems are described in detail in Chapter 6.

3.5.3 Microwave applicator

The microwave applicator is responsible for delivering the microwave power to the exposed surface of concrete. In microwave-assisted selective

Figure 3.12 Microwave applicators: (a) antenna (horn); (b) waveguides.

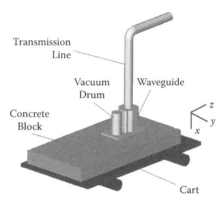

Figure 3.13 Schematic illustration of transmission line and applicator in a typical microwave-assisted selective demolition tool.

demolition tools, the applicator is usually either a simple waveguide or an antenna used to spread or localise the microwave power on the surface of concrete (Figure 3.12). The applicator design and size vary with the microwave frequency range used by the demolition tool. Furthermore, the applicator should include sufficient safety features to prevent leakage of the reflected microwave power and to contain it to avoid health hazards to the operators. In a number of the proposed microwave-assisted demolition tools, the applicators are also equipped with an attached vacuum system to collect the concrete debris produced during the demolition process to reduce health hazards caused by dust generation (Figure 3.13). The design basics for the applicators for typical microwave-assisted selective demolition tools are discussed in detail in Chapter 6.

3.5.4 Attached mechanical demolition tools (optional)

The simultaneous use of mechanical tools and microwave heating has been suggested by a number of inventors to improve selective demolition efficiency and energy savings and for reducing associated emissions. It has been debated that microwave heating may result in softening of the concrete so that it can be removed using considerably smaller mechanical forces.

3.5.5 Control unit

The control unit is the heart of the microwave-assisted demolition system and is responsible for monitoring and controlling the demolition process based on the information provided by the user as well as the feedback from a network of sensors used to monitor the performance of the various components. The main responsibilities of the control unit include selecting the appropriate microwave frequency, microwave power, and heating duration to achieve the desired depth and rate of demolition, monitoring the performance of various components in the system to ensure optimal performance, monitoring the safety of the operator by monitoring the amount of possible leakage, and identifying potential faults in the system. In automated microwave-assisted selective demolition systems, the control unit is also responsible for controlling the movement of the applicator.

3.5.6 Robotic arm

To improve the safety of operators, extending potential working hours, and eliminating the need for training and hiring of skilled operators, microwave-assisted systems could be designed to operate automatically by including remote operational and robotic features. A robotic arm is basically a mechanical arm that can be programmed to move the microwave applicator along a predefined route. The control of the robotic arm requires the use of navigation systems such as laser scanners for locating of obstacles and walls and for measuring the demolition depth achieved before moving the applicator is moved to the next location.

3.6 WORKING PRINCIPLES OF MICROWAVE-ASSISTED SELECTIVE DEMOLITION OF CONCRETE

A number of examples of successful experiments verifying the capabilities of microwave heating for potential uses in the demolition of concrete were reviewed in the previous sections. This section discusses the working principles of the microwave-assisted selective demolition method and explains the phenomena leading to the development of forces big enough to break up materials such as concrete and rocks.

3.6.1 Main causes of concrete demolition

The results of various analytical and experimental studies have confirmed that the use of high ISM microwave frequencies can result in the selective demolition of the microwave-exposed surface of concrete. The results of such studies suggest that demolition of the microwave-exposed concrete area occurs due to a combination of two phenomena: differential thermal heating and pore pressure development (Figure 3.14). Acquiring a thorough understanding of the underlying phenomena leading to the selective removal of concrete using microwaves is essential for proper use of the microwave-assisted concrete demolition equipment and for fine-tuning the entire demolition process.

3.6.2 Differential thermal heating

As shown in Chapter 1, the penetration depth of microwaves in concrete is basically a function of two parameters: the dielectric properties of concrete and microwave frequency. The dielectric properties of concrete are themselves functions of various other parameters, including the concrete composition, proportions of concrete constituents, water content of concrete, and microwave frequency (Figure 3.15).

It has been shown that, among these factors, the water content of concrete plays the most important role in determining the amount of the microwave energy absorbed by and dissipated in the concrete. The higher the water

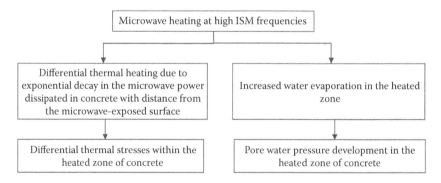

Figure 3.14 The main causes of concrete demolition in microwave-assisted selective demolition techniques.

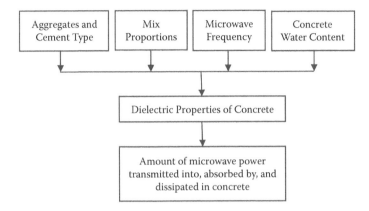

Figure 3.15 Factors determining the dielectric properties of the concrete.

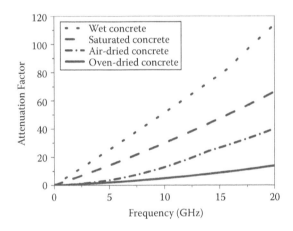

Figure 3.16 The variation of attenuation factor with concrete water content and microwave frequency. The attenuation factor is directly proportional to the
microwave energy dissipated in concrete; thus, the variation in the attenuation factor may represent the variation in the microwave power dissipated
in concrete.

content of concrete, the faster the microwave energy would be absorbed
by and dissipated in the concrete (Figure 3.16). Furthermore, the reflection
coefficient of concrete is directly related to its water content. The higher
the water content of concrete, the higher the reflection coefficient would
be, and thus a higher proportion of the incident microwave energy may be
reflected from the concrete surface (Figure 3.17). Apart from the dielectric
properties, the behaviour of microwaves in concrete is also a function of
microwave frequency. For a given concrete, there is an inverse relationship

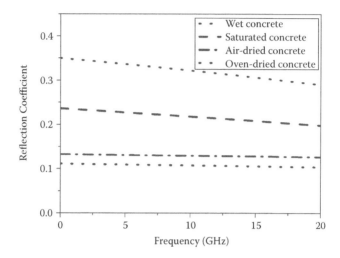

Figure 3.17 The variation in the reflection coefficient of concrete with concrete water content and microwave frequency.

between the penetration depth of microwaves in concrete and microwave frequency. This means that the higher the microwave frequency, the less the penetration depth would be. Therefore, at higher ISM microwave frequencies, the microwave energy is expected to be dissipated within the surface layer of the concrete rather than penetrating deep into the concrete specimen; the higher the microwave frequencies, the thinner this affected surface layer would be. In microwave-assisted selective demolition of concrete, such relationships between the penetration depth of microwave in concrete and the concrete's water content and microwave frequency are used to concentrate the energy within a selected area of the concrete component, thereby causing local surface delamination.

The delamination of the heated zone of the concrete has been shown to be caused by a combination of two forces: differential thermal stresses and elevated pore water pressure. The differential thermal stresses are a direct outcome of the differential heating of the heated parts of the concrete caused by the relatively fast decay of microwave power within the concrete surface layer, especially at higher ISM microwave frequencies. As a result of the confinement of the energy within a thin surface layer of concrete, microwave heating at higher ISM microwave frequencies generates high-temperature gradients inside the concrete, between the microwave-exposed surface and the cooler interior. Figures 3.18 to 3.21 show the variation in temperature within a 10-cm (cube) concrete specimen subjected to microwaves of three different ISM frequencies (i.e., 2.45, 10.6, and 18 GHz). The results presented in these figures have been obtained by simulating the microwave-concrete behaviour using COMSOL Multiphysics

Table 3.1 Mechanical and thermal properties of concrete

Property	Assumed values	Unit
Density	2300 (143.58)	Kg/m^3, (lb/ft^3)
Specific heat	1000 (0.24)	$J/[Kg°K]$, $(Btu/[lb.°F])$
Thermal conductivity	3 (1.73)	$W/[m.K]$, $(Btu/[ft.hr.°F])$
Thermal expansion coefficient	12×10^{-6} (6.67×10^{-6})	$1/°C$, $(1/°F)$
Heat transfer coefficient of heat flux	10.0 (1.76)	$W/[m^2,°K]$, $(Btu/[ft^2.hr°F])$
Modulus of elasticity	48.5 (7034.33)	GPa (ksi)
Poisson's ratio	0.12	
Surface emissivity	0.9	
Initial temperature	25 (77)	$°C$, $(°F)$

software, a commercially available finite-element simulation package. The dielectric properties as well as the thermal and mechanical properties of the concrete used to achieve these results are summarised in Table 3.1.

The microwave power was assumed to be constant at 1.1 MW/m². Throughout the present chapter, this example of microwave heating of a typical concrete specimen is used to illustrate the behaviour of microwaves in concrete. Furthermore, to understand the importance of the water content of concrete in the context of the microwave-assisted demolition process, four different concrete water contents are considered and discussed. These include the oven-dried condition, representing a concrete that has lost all its free water after being completely dried in an oven; the air-dried condition, representing a concrete that has partially lost its free water through exposure to ambient conditions for a relatively long period; the saturated condition, representing a concrete with completely saturated pores but with a dry surface; and the wet condition, representing a concrete with saturated pores and a thin layer of water standing on its surface.

Figure 3.18 shows the temperature distribution across the incident surface of the concrete as well as the penetration depth of microwaves in the concrete specimen for a saturated 10-cm thick concrete specimen subjected to microwaves of three different frequencies at a constant microwave power of 1.1 MW/m². This figure clearly shows that the penetration depth decreases with an increase in the microwave frequency. Furthermore, it can be seen that the microwave-exposed area of the concrete tends to be smaller at higher frequencies than lower ISM frequencies. This is due mainly to the smaller cross section of the waveguides used for transferring the microwave energy at higher microwave frequencies. The size of the standard waveguides used to transfer microwaves at different microwave frequencies is listed in Table 3.2.

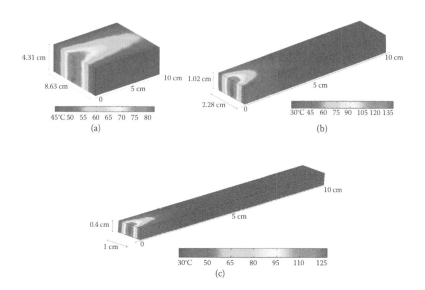

Figure 3.18 Temperature distribution across the heated zone of a saturated concrete after (a) 5 seconds of microwave heating at 2.45-GHz frequency; (b) 2 seconds of microwave heating at 10.6-GHz frequency; and (c) 1 second of microwave heating at 18-GHz frequency (incident power = 1.1 MW/m2).

Table 3.2 Dimensions of standard waveguides

Microwave frequency (GHz)	Designation	Width (mm)	Height (mm)
2.45	WR340	86.36	43.18
10.6	WR90	22.86	10.16
18	WR42	10.66	4.31

Figures 3.19 to 3.21 show that, regardless of the microwave frequency and concrete water content, the temperature in the concrete decays exponentially with distance from the exposed surface. Again, it is also obvious that at a constant microwave power, there is an inverse relationship between the penetration depth of microwaves with the water content and the microwave frequency. As a result, an increase in either the water content of concrete or the microwave frequency results in faster decay in the microwave power within the concrete specimen.

Figure 3.22 shows the temperature distribution across the microwave incident surface of the concrete. The temperature distribution across the concrete surface is a function of the microwave mode excited by the microwave generator and the waveguide used to deliver the power. The modes of transmission in the waveguides are basically the multiple solutions of Maxwell's equations governing the behaviour of the electromagnetic waves

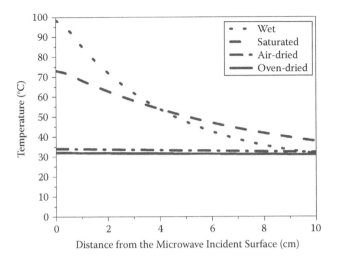

Figure 3.19 Temperature distribution in concrete after 5 seconds of microwave heating at 2.45-GHz frequency.

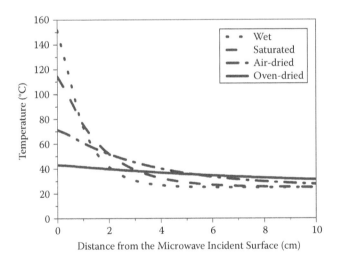

Figure 3.20 Temperature distribution in concrete after 2 seconds of microwave heating at 10.6-GHz frequency.

in various media. Modes are eigenvalues of the equation system (Maxwell's equation and boundary conditions). Each microwave mode is character-ised by a cutoff frequency, which is the minimum frequency at which the mode can exit the waveguide. If the frequency of the excited mode is above the cutoff frequency, electromagnetic energy will be transmitted down

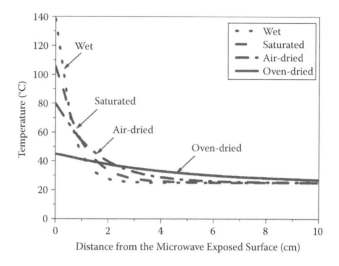

Figure 3.21 Temperature distribution in concrete after 1 second of microwave heating at 18-GHz frequency.

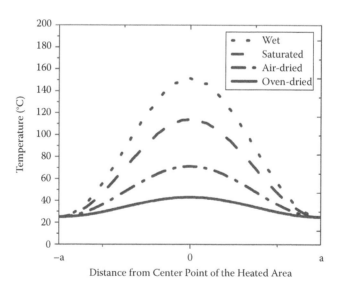

Figure 3.22 Temperature distribution across the microwave incident surface of concrete (waveguide width of 2 × a) after 2 seconds of microwave heating at 10.6-GHz frequency.

the guide with little attenuation. If it is below the cutoff frequency, the wave is described as evanescent and will be attenuated strongly within a short distance.

For a homogeneously filled waveguide, the cutoff frequency, f_c, is given by

$$f_c = \frac{1}{2\pi\sqrt{\mu_0\varepsilon_0\varepsilon}\sqrt{\left(\frac{m\pi}{a}\right)^2 + \left(\frac{n\pi}{b}\right)^2}}$$

where a and b are the dimensions of the waveguide, and m and n are the mode indices. The cutoff frequency characteristic of a waveguide must be used to choose the appropriate waveguide and horn for a specific frequency and microwave mode.

The propagation modes in the waveguides vary with the polarisation, shape, and size of the waveguide. The longitudinal mode is a particular form of standing wave pattern developed by waves confined in the cavity. On the other hand, the transverse modes include transverse electric (TE) modes, transverse magnetic (TM) modes, and transverse electromagnetic (TEM) modes. TE and TM modes have, respectively, no electric field and no magnetic field in the direction of propagation. TEM modes have neither electric nor magnetic field in the direction of propagation. Finally, hybrid modes are the propagation modes with both electric and magnetic field components in the direction of propagation. In the hollow waveguides used in the transmission lines of microwave-assisted demolition tools, TEM modes are not possible. This is because obtaining the TEM modes as the solution of Maxwell's equations requires the electric field to have zero divergence and zero curl and be equal at the boundaries, requiring it to be equal to zero. Because of the boundary conditions, the waves that propagate inside a homogeneously filled waveguide are different from the TEM waves and are known as TE waves and TM waves. TE waves have $E_x = 0$; TM waves have $H_x = 0$. For TE waves, there is a component of H along the direction of propagation; for TM waves, it is the E component that exists in the same direction. In both cases, energy is carried by the electric and magnetic fields associated with the wave. It is possible to have several modes of propagation inside a particular waveguide. The mode that has the lowest frequency in a particular waveguide is known as the dominant mode. Generally, the waveguide used for microwave heating has dimensions such that only the dominant mode propagates along it. The fundamental TE and TM modes, TE_{10} and TM_{11}, respectively, are usually the two most dominant modes [16]. The TE_{10} mode is the dominant mode excited by most of the basic microwave heating systems available today. Field lines for the TE_{10} mode in a rectangular waveguide are shown in Figure 3.23. In the example problem used in the present chapter, the rectangular waveguides used are

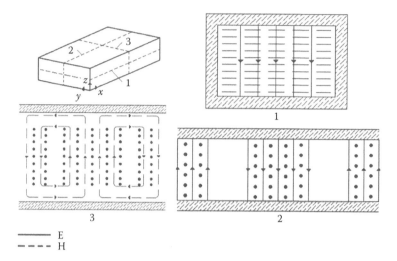

Figure 3.23 Field lines for the TE10 mode in a rectangular waveguide.

assumed to excite the TE_{10} mode. As seen in Figure 3.22, the sinusoidal shape of the TE_{10} mode results in a sine² shape temperature distribution across the microwave incident surface.

As illustrated through the example (Figures 3.18 to 3.22), the exponential decay in the microwave power within concrete, especially at high concrete water contents or high microwave frequencies, may result in considerably high temperature gradients within the concrete's surface layer. Such nonuniform heating within a very short time leads to a high differential temperature gradient and thus high thermal stresses. Figures 3.24 to 3.26 show the thermal stresses developed in the concrete specimens subjected to microwaves at different frequencies and at a constant microwave power of 1.1 MW/cm². Furthermore, the stress developed across the microwave incident surface of the concrete is plotted in Figure 3.27. Such plots may be of use when choosing the appropriate microwave specifications for a given nominal strength and depth of spalling. Similarly, for a specific incident microwave power, microwave frequency, and concrete water content, the approximate spalling depth can be easily estimated. As seen in Figures 3.24 to 3.26, considerably high differential thermal stresses may be developed within a concrete specimen after just a few seconds of microwave heating, especially at high frequencies and water contents, at a typical microwave power used in industrial heating. Such stresses are commonly in the form of radial compressive stresses and may result in delamination of the surface layer of the concrete when the compressive strength of concrete is exceeded under confined conditions.

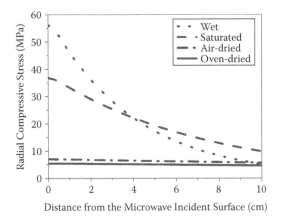

Figure 3.24 Radial compressive stress in concrete after 5 seconds of microwave heating at 2.45-GHz frequency.

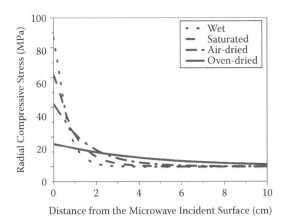

Figure 3.25 Radial compressive stress in concrete after 2 seconds of microwave heating at 10.6-GHz frequency.

It is interesting to note that, for a given microwave power and heating period, higher microwave frequencies result in higher stresses being generated within a thinner surface layer of the concrete. The effect of microwave frequency on the maximum thermal stresses generated in concrete specimens with different water contents is shown in Figure 3.28. As seen, an increase in microwave frequency can lead to significantly higher compressive stresses.

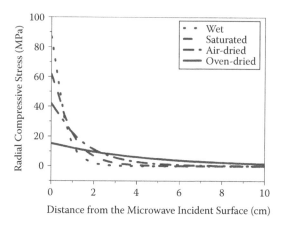

Figure 3.26 Radial compressive stress in concrete after 1 second of microwave heating at 18-GHz frequency.

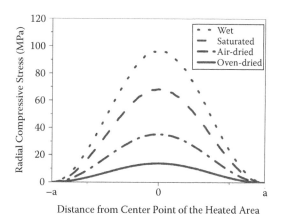

Figure 3.27 Radial compressive stress distribution across the microwave incident surface of concrete after 2 seconds of microwave heating at 10.6-GHz frequency.

Another important effect that should be taken into consideration to understand the behaviour of microwaves in concrete and the magnitude of the thermal stresses in concrete is the significant effect water content has. As seen in Figures 3.24 to 3.28, a higher water content leads to faster and more concentrated stress development in the heated concrete specimen, which speeds up the surface delamination process. In practice, most of the contaminated concrete blocks are old and are likely to be in an air-dried condition; hence, drenching the concrete surface may help both to speed up the process and to control spalling depth.

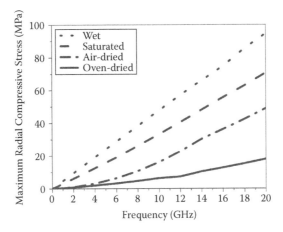

Figure 3.28 The variation of maximum compressive stress in concrete with frequency after 1 second of microwave heating.

3.6.3 Pore water pressure

Apart from the thermal stresses developed because of differential heating of the concrete surface at higher ISM microwave frequencies, the development of relatively high pore water pressure within the heated surface layer of the concrete is considered the second important factor causing delamination of the microwave-exposed surface.

Water is a key ingredient in the manufacture of concrete, and concrete pores are filled partially with water and air. The main source of moisture in concrete is the mixing water used to make the concrete. Water in concrete can also stem from other sources, including the water used for wet curing and on exposure to the weather. Mixing water is added to concrete during batching to provide the moisture required for the hydration of cement and to achieve the workability required in placing and finishing the concrete. Although the majority of this water is used up by the hydration process or is lost through bleeding and evaporation, a small fraction of this water remains within the pores of the concrete. Wet curing is another source of water in the concrete. Wet curing basically aims to ensure that the hydration process continues by providing external sources of water so that the target strength and other requirements are achieved.

The total amount of moisture in concrete, in the form of water or water vapour, is expressed as the moisture content. Moisture content is usually expressed as a percentage of the mass of the concrete. This moisture, in the form of water or water vapour, is stored in the various pores within the concrete. The pores in concrete include gel pores, which are micropores with radii ranging from 0.5 to 10 nm; capillary pores, which are mesopores with radii ranging from 5 to 5000 nm; macropores, which are basically caused by deliberately entrained air or inadequate compaction; and finally, pores caused by the cracks at the aggregate-mortar interface as a result of shrinkage. The moisture stored in concrete pores may be in the form of the water or water vapour. The relative humidity in concrete varies with time as water vapour moves in or out of the concrete to establish equilibrium with the changing ambient exposure conditions. Unless the ambient relative humidity is considerably low, the relative humidity in concrete remains high even after the majority of water has evaporated (75%).

When concrete is exposed to microwaves, as a result of dielectric losses, microwaves penetrating the concrete act as a volumetrically distributed heat source. Water in the concrete is a strong dipole and is easily heated, as it absorbs the microwave energy. As a result, the water within the concrete evaporates. When the evaporation rate exceeds the vapour migration rate, pore pressure builds.

Figures 3.29 and 3.30 show the pore pressure developed in a representative saturated concrete specimen subjected to microwaves at two different frequencies (10.6 and 18 GHz, respectively). As seen in these figures, for a typical concrete specimen heated at a typical industrial microwave power level (1.1 MW/m^2), the pore water pressure generated in the concrete can easily exceed the tensile strength of typical concretes within a few seconds of heating, thus leading to delamination of the concrete surface. Again, similar to the trend observed for the temperature distribution and the thermal stresses distribution, pore pressure decays exponentially with distance from the exposed surface. Furthermore, the decay rate of the pore pressure within the concrete seems to be directly related to the microwave frequency;

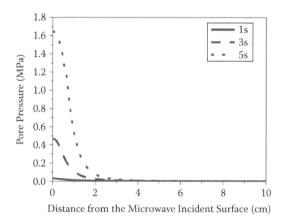

Figure 3.29 Pore pressure in saturated concrete after 5 seconds of microwave heating at 10.6-GHz frequency.

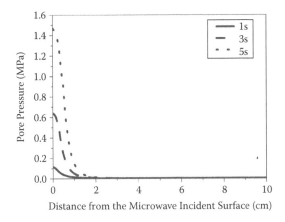

Figure 3.30 Pore pressure in saturated concrete after 3 seconds of microwave heating at 18-GHz frequency.

that is, the higher the microwave frequency, the higher the pore pressures developed in a thinner concrete surface layer.

Comparing the magnitudes of the pore pressure and the compressive thermal stresses developed in the microwave-exposed surface of the concrete highlights the prominent role that thermal stresses play in the delamination of the microwave-exposed concrete. Nevertheless, considering the inherent weakness of concrete in tension, the contribution of pore pressure should also be considered in making predictions and fine-tuning the microwave-assisted selective demolition methods to achieve the desired demolition outcome.

3.6.4 Effects of the reinforcing bars on the microwave-assisted removal efficiency

Although strong in compression, concrete is relatively weak in tension and thus requires reinforcement to carry tensile forces (Figure 3.31a). Also, most concrete structural elements will be subjected to tensile stresses during service. Without adequate reinforcement, concrete elements are highly prone to cracking and even failure. In addition, even elements that are not subjected to tensile forces directly are likely to experience tensile stresses caused by shrinkage and creep. To this end, reinforcements, including steel reinforcing bars, steel fibres, glass fibres, or plastic fibres, are usually embedded or added to resist the tensile effects present. Steel reinforcing bars are the most commonly used type of reinforcement. They are detailed to resist tensile stresses at critical regions of the concrete to avoid unacceptable cracking or structural failure (Figure 3.31b).

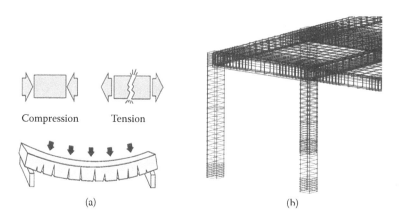

Compression   Tension

(a)   (b)

Figure 3.31 (a) Concrete is strong in compression but weak in tension. (b) Typical steel reinforcement bars in RC structural elements.

Understanding the effect the embedded reinforcing bars have on the distribution of microwave power in reinforced concrete is crucial to achieve a realistic estimate of the thermal stresses and pore water pressure developed during the microwave selective demolition process. Contrary to concrete, which absorbs the microwave energy partially, the steel rebars within concrete reflect the majority of the incident microwave power. As a result of this reflection back toward the microwave-exposed surface of concrete, the microwave field intensity within the cover concrete above the reinforcing bars near the surface is increased (Figure 3.32). This is a positive phenomenon if the aim of the selective concrete demolition technique is to remove the surface layer of the cover concrete in applications such as decommissioning of nuclear power plants and can result in increased performance, translating into cost and energy savings provided the desired depth of surface removal is less than the thickness of the concrete cover.

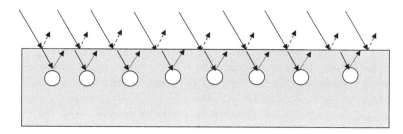

Figure 3.32 Reinforcing bars in concrete reflect the microwave power back toward the microwave-exposed surface of concrete, increasing the amount of the dissipated microwave power within the cover zone.

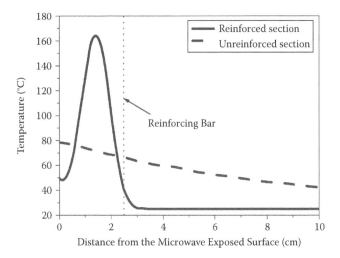

Figure 3.33 Temperature distribution in reinforced saturated concrete after 5 seconds of microwave heating at 2.45-GHz frequency and 1.1-MW/m2 incident power.

Figures 3.33 to 3.35 show the temperature distribution within reinforced concrete at three different ISM microwave frequencies estimated through numerical simulation of the representative concrete considered in the present chapter. The presence of reinforcing steel bars in concrete usually adds to the complexity of the analytical models for the estimation of the distribution of microwaves in concrete. A number of assumptions have been suggested by various researchers to simplify the interaction between the microwaves and embedded steel rebars to estimate the energy and thus temperature distribution. The methodology used in such studies as well as the methodology for estimation of the microwave behaviour in steel-reinforced concrete is discussed in Chapter 6.

As seen in Figures 3.33 to 3.35, as a result of the increase in microwave power intensity in the cover concrete caused by reflection by the reinforcing bars, the presence of reinforcing bars leads generally to a higher temperature rise within the surface layer of concrete. The effect of the presence of reinforcing bars on the microwave-assisted demolition process is expected to decrease with a decrease in the microwave penetration depth. The effect is especially minimal when the microwave penetration depth is less than the concrete cover to the embedded reinforcing steel (higher ISM microwave frequencies). The microwave penetration depth may decrease with an increase in microwave frequency or water content of concrete. As can be seen in Figures 3.33 and 3.35, the effect of the presence of reinforcing bars on the temperature distribution generated in saturated concrete for microwave frequencies of 10.6 and 18 GHz is almost negligible.

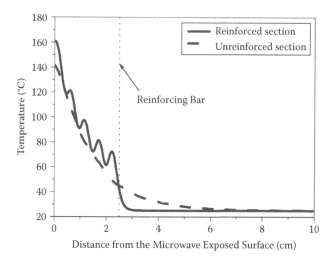

Figure 3.34 Temperature distribution in reinforced saturated concrete after 2 seconds of microwave heating at 10.6-GHz frequency and 1.1-MW/m2 incident power.

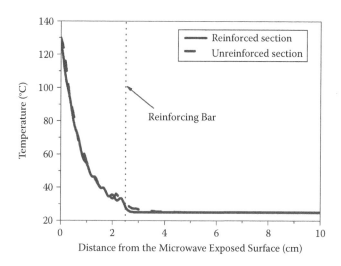

Figure 3.35 Temperature distribution in reinforced saturated concrete after 1 second of microwave heating at 18-GHz frequency and 1.1-MW/m2 incident power.

3.6.5 Fine-tuning of microwave-assisted selective concrete removal process

As discussed, differential thermal stresses and pore water pressure development are the main causes of concrete demolition in microwave-assisted selective demolition tools. As demonstrated through an illustrative example of a typical concrete specimen subjected to microwaves of different frequencies, the magnitude and distribution of thermal stresses and pore pressure of concrete are a function of the concrete dielectric properties, microwave frequency, and applicator design. On the other hand, the dielectric properties of concrete depend on concrete mix proportions, type of cement and aggregates used, and most important the water content of the concrete. The water content was shown to play a crucial role in determining the magnitude and distribution of thermal stresses and pore water pressure developed in the concrete exposed to microwaves. Furthermore, the presence of embedded steel reinforcement in the concrete was also shown to affect the microwave-assisted selective demolition process by affecting the distribution of microwave power in the microwave-exposed zone of the concrete.

Keeping the previous discussion in mind, it is obvious that achieving a particular demolition thickness and rate requires selection of an appropriate range of microwave frequency and microwave power based on the properties of the concrete and the applicator design. On the other hand, concrete is a nonhomogeneous material, and the properties of concrete, especially water content, could vary significantly from one section of the concrete specimen to another because of various factors, including exposure to different environments. Therefore, optimising the microwave-assisted selective demolition process to maximize performance and to minimise costs and energy use could be a relatively complicated process requiring the use of sophisticated control systems. The recommended framework for the selection of appropriate microwave frequency, power, and heating duration to achieve a particular demolition rate and thickness using a typical microwave-assisted concrete selective demolition tool is summarised in Figure 3.36.

First, considering the important effect of water content, the selected section of concrete should be drenched with water shortly prior to the start of demolition (recommended time 5 to 10 minutes, but this depends on the permeability of the concrete) to improve the removal efficiency of microwave-assisted demolition. The minimum drenching duration to saturate the concrete to the desired depth from the surface (approximately equal to the desired removal thickness) could be estimated using the basic properties of the concrete. The basic mechanical properties of the concrete, including permeability, permittivity, and tensile and compressive strengths, should either be determined or estimated from available structural records and information models or measured using available nondestructive testing

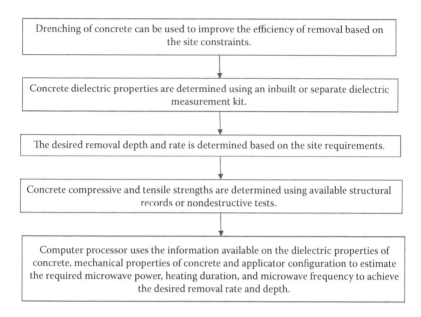

Figure 3.36 Proposed framework for selection of appropriate microwave frequency, microwave power, and heating duration.

methods. Such information should then be stored in the control system using the user interface. It must be noted that drenching of the selected zones of the concrete component with water may not always be possible because of site restrictions and leakage of water. Second, an inbuilt dielectric measurement kit or network analyser can be used to estimate the dielectric properties of the concrete. Third, the processor unit of the control system makes use of the estimated dielectric properties, mechanical properties, applicator configuration information, and other required data imported from the various databases to estimate the required microwave frequency, microwave power, and heating duration to achieve thermal stress and pore pressure levels sufficient to exceed the compressive or tensile strengths of the concrete, respectively. The procedure used by the control system mentioned is described in detail in Chapter 5.

3.7 MICROWAVE-ASSISTED DRILLING OF CONCRETE

One variation of the microwave-assisted selective demolition method is the microwave-assisted drilling method for concrete. Drilling into concrete is an inevitable part of most construction projects. Currently, mechanical drills drill holes of different sizes in concrete; these holes can be used for installing utility wiring and piping, attaching various equipment and supporting

Figure 3.37 Microwave drill invented by Jerby and Dikhtyar. (2000) [18]. (Courtesy of Professor Eli Jerby, faculty of engineering, Tel Aviv University.)

elements, or installing shear connectors for better connection between the old and new concrete. However, mechanical drilling produces considerable noise, dust, and vibration, posing serious health hazards to drill operators and so on. A novel microwave-assisted method has been proposed by researchers at the University of Tel Aviv in Israel as an alternative to conventional drilling (Figure 3.37). Jerby and Dikhtyar [17,18], the inventors of this technique, have reported that the microwave drill can make a 2-mm diameter, 2-cm deep hole in concrete in less than a minute, rendering it competitive vis-à-vis conventional drilling methods in terms of performance while eliminating most of the drawbacks of conventional methods.

In a microwave-assisted drill, the microwave energy should be concentrated at a very small spot, normally much smaller than the microwave wavelength itself. This is because most holes to be drilled will range in the order of a few millimetres to a few centimetres in diameter. In a microwave drill, this is done using a near-field microwave concentrator. As shown in Figure 3.38, although the microwave-assisted selective demolition technique discussed previously makes use of an antenna to localise (spread, in some cases) microwave power to the desired surface area of concrete, the microwave-assisted drill is designed to localize the microwave power into a considerably smaller area (a point). The applicator unit in microwave-assisted drills consists of an open-ended coaxial waveguide with a movable centre electrode that concentrates the microwave power into a specific point. The concentrator itself is then inserted into the molten hot spot to make the hole. Once the hole boundaries are shaped, the concentrator is extracted from the hole and the concrete cools into a newly shaped hole. The materials for the coaxial waveguide and the movable center electrode should be selected from materials able to withstand the high temperatures generated.

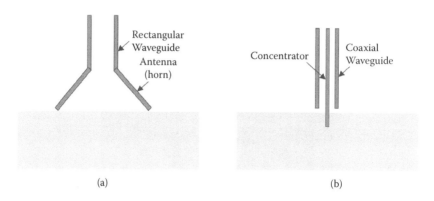

Figure 3.38 Schematic illustration of the applicator in a (a) microwave-assisted selective demolition tool and a (b) microwave drill.

Apart from the microwave applicator, other main components of microwave-assisted drills are similar to those used in microwave-assisted selective demolition tools discussed previously and include a microwave source and power supply, an isolator to prevent the device from potential damage caused by the reflected power, a tuner to optimise the power absorbed by the concrete, and a microwave transmission line to transmit the microwave power to the applicator.

As the microwave-assisted drill progresses in the concrete, the debris is densified to the wall or converted to a glossy material [17]. The glossy materials attached to the hole surfaces are soft and can be easily removed mechanically. The size of the hole created by the microwave-assisted drill can be adjusted by varying the dimensions of the concentrator and the coaxial waveguide or through further microwave heating of the boundary walls. The hole's depth can be increased by applying successive cycles of microwave drilling and mechanical removal of the glossy debris from the hole surfaces. Figure 3.39 shows a 13-mm diameter, 10-cm depth hole made in four cycles of microwave drilling in a concrete slab using the microwave drill tool invented by Jerby and Dikhtyar [17].

Although the microwave-assisted selective demolition tools discussed in the previous sections rely on development of high-differential thermal stresses and pore water pressure within the microwave-exposed region of the concrete to cause delamination, the microwave-assisted drill relies mainly on the thermal runway effect developed by concentrating and focussing the microwave energy at a considerably smaller area on the surface of concrete to "melt" the concrete at the heated spot.

Apart from concrete, the microwave drill is suitable for ceramics and glass as well as the other dielectric materials with good microwave absorbent properties. Inventors of this technique believe that, apart from the

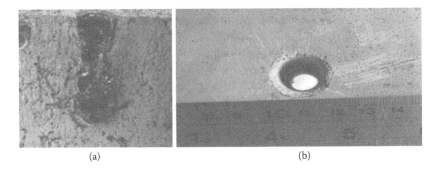

(a)   (b)

Figure 3.39 (a) A cross-sectional cut of a microwave-drilled concrete slab. (b) A 13-mm diameter and 10-cm deep hole made in a concrete slab using the microwave drill invented by Jerby and Dikhtyar (2000) [18]. (Courtesy of Professor Eli Jerby, faculty of engineering, Tel Aviv University.)

construction and renovation industries, the microwave drill can be used in production lines in the electronics industry and for ceramics, glass, and so on to improve efficiency and to reduce health hazards associated with conventional drilling.

Similar to other microwave-assisted selective demolition tools, the main advantages of the microwave drill compared to conventional drills include reducing health hazards and disturbance caused by dust and noise generated during drilling.

3.8 SUMMARY

Conventional concrete demolition and drilling techniques usually generate a lot of noise and dust. Furthermore, a major problem associated with state-of-the-art concrete demolition technique is the lack of selectivity. Concrete is a relatively tough material; therefore, it is difficult to drill into or demolish a part of a concrete component without affecting/damaging the adjacent sections. A selective concrete demolition technique able to demolish precisely a specific part of a concrete component, without causing damage to adjacent sections or surface finishes, is sorely needed in today's repair and retrofitting applications. After reviewing the current state-of-the-art selective concrete demolition techniques, this chapter introduced a novel highly selective concrete demolition technique based on microwave heating at higher ISM frequencies.

Microwave-assisted selective demolition makes use of microwave heating at higher ISM frequencies (>2.45 GHz) to concentrate microwave power within a thin surface layer of the concrete, thereby generating a region of high thermal stress and pore pressure differential between the microwave-heated surface layer and the cooler concrete in the interior. Spalling of the concrete surface occurs when the pore pressure or thermal stresses exceed the concrete tensile or compressive strengths, respectively. The microwave-assisted demolition of concrete is highly selective, enabling delamination to be precisely controlled, and has been reported to generate significantly less noise and dust than conventional techniques.

This chapter also introduced another variation of the microwave-assisted selective demolition technique for use in drilling of concrete. The microwave drilling technique involves the continuous and localised microwave heating of concrete at one precise location, leading to holes being "drilled" into the concrete with relatively minor noise and dust generation. The phenomena leading to removal of concrete surfaces when heated with microwaves as well as the working principles of microwave assisted demolition and microwave drilling techniques were described. Furthermore, the effects of various factors, including microwave frequency, microwave power, and water content of concrete, on the efficiency of these methods were discussed in detail.

REFERENCES

1. Spalding, B., Volatility and extractability of strontium-85, cesium-134, cobalt-57, and uranium after heating hardened Portland cement paste. *Environmental Science Technology*, 2000, **34**:5051–5058.
2. Dickerson, K.S., Wilson-Nichols, M.J., and Morris, M.I., *Contaminated Concrete: Occurrence and Emerging Technologies for DOE Decontamination.* DOE-ORO-20341995. Oak Ridge, TN: US Department of Energy, National Laboratory.
3. Ayers, K.W., *Feasibility of Recycling Contaminated Concrete.* Nashville, TN: Vanderbilt University, 1998.
4. Akbarnezhad, A., and Ong, K.C.G., Microwave decontamination of concrete. *Magazine of Concrete Research*, 2010, **62**(12):879–885.
5. Akbarnezhad, A. and Ong, K.C.G., Thermal stress and pore pressure development in microwave heated concrete. *Computers and Concrete*, 2011, 8(4):425–443.
6. Ong, K.C.G. and Akbarnezhad, A., Thermal stresses in microwave heating of concrete. In *Proceedings of Our World in Concrete and Structures*, Singapore, August 16–17, 2006, 297–310.
7. Zi, G. and Bažant, Z.P., Decontamination of radionuclides from concrete by microwave heating. II: Computations. *Journal of Engineering Mechanics*, 2003, **129**(7):785–792.
8. Bažant, Z.P. and Zi, G., Decontamination of radionuclides from concrete by microwave heating. I: Theory. *Journal of Engineering Mechanics*, 2003, **129**(7):777–784.
9. Watson, A., Breaking of concrete. In *Microwave Power Engineering*, Vol. 2, E.C. Okress, ed. New York: Academic, 1968, 108–118.

10. Watson, A., Methods of cracking structures and apparatus for cracking structures, US Patent 3430021, filed May 2, 1966, and issued February 25, 1969.
11. White, T.L., Bigelow, T.S., et al., Mobile system for microwave removal of concrete surfaces, U.S. Patent 5635143 A, filed September 30, 1994, and issued June 30, 1997.
12. White, T.L., Grubb, R.G., Pugh L.P., Foster, D., Jr., and Box, W.D., Removal of contaminated concrete surface by microwave heating—phase 1 results. In *Proceedings of 18th American Nuclear Society Symposium on Waste Management*, Tucson, AZ, March 1–5, 1992. Available at   HYPERLINK "http://inis.iaea.org/search/search.aspx?orig_q=RN:23044093" http://inis.iaea.org/search/search.aspx?orig_q=RN:23044093
13. White, T.L., Foster, D., Jr., Wilson, C.T., and Schaich, C.R., *Phase II Microwave Concrete Decontamination Results*. ORNL Rep. No. DE-AC05-84OR21400. Oak Ridge, TN: National Laboratory, 1995.
14. Puschner, H.A., Apparatus for heating material by means of microwave device, US Patent 3443051 A, filed July 25, 1966, and issued May 6, 1969.
15. Lindroth, D.P., Morrell, R.J., and Blair, J.R., Microwave assisted hard rock cutting, US Patent 5003144 A, filed April 9, 1990, and issued March 26, 1991.
16. Balanis, C.A., *Advanced Engineering Electromagnetics*. New York: Wiley, 1989.
17. Jerby, E. and V. Dikhtyar, Drilling into hard non-conductive materials by localized microwave radiation. In *Advances in Microwave and Radio Frequency Processing*, M. Willert-Porada (Ed.), Berlin Heidelberg: Springer, 2006, 687–694.
18. Jerby, E. and Dikhtyar, V., Method and device for drilling, cutting, nailing and joining solid non-destructive materials using microwave radiation, US Patent 6114676, filed January 19, 1999 and issued September 5, 2000.

Chapter 4

Microwave-assisted concrete recycling

4.1 INTRODUCTION

The construction industry is the largest consumer of raw materials worldwide, with an estimated consumption of up to 50% of all materials. As a result of the enormous amount of energy consumed in the production and transportation of materials together with the energy consumed by construction operations and the operation of buildings, the global construction industry has been estimated to be responsible for up to 40% of the total energy use and up to 30% of the total associated carbon emissions. In addition, up to 50% of the total waste generated in some countries has been estimated to be directly or indirectly related to the construction industry [1,2].

The substantial amount of materials used by the construction industry contributes to some of the biggest environmental concerns. The major environmental concerns related to construction materials include (1) upstream impacts related to the depletion of nonrenewable natural resources as well as the energy used and the waste generated during the production of materials; and (2) downstream impacts concerned with the huge amount of material waste generated at the end of the service life of the materials as well as the energy and associated carbon emissions incurred during processing of demolition wastes. Dealing with the waste material generated during the production of construction materials as well as during the construction and demolition (C&D) phases of buildings and other structures has been a major focus of research over the past few decades. The United States alone produces around 140 tons of C&D waste each year, accounting for almost 29% of the total solid waste generated in the country [3]. Europe is estimated to produce about 970 tons of C&D waste annually [4]. In Australia, about 14 million tons of solid wastes are dumped in landfills annually, and 44% of this total is estimated to be attributed to the construction industry [5].

The most common way of dealing with the huge amount of C&D waste generated every day is to invest waste with new usable life (as the same or a different material) through recycling. The benefits of effective C&D waste recycling are economic, environmental, and social. Recycling reduces significantly the environmental impacts of construction by reducing the space needed as landfill for waste disposal and reducing the demand for raw materials through providing alternative sources of construction materials. As a result, the environmental impacts caused by the extraction of nonrenewable materials, including extensive deforestation, top soil loss, air pollution, and pollution of water reserves can be mitigated to some extent. Recycling of C&D waste can also bring about economic and social benefits related to the creation of new jobs. According to the 2002 report of the US Environment Agency (EPA) the incineration or landfilling of 10,000 tonnes of waste may create 1 job and 6 jobs, respectively, while recycling the same amount of waste may create 36 jobs [6]. The market for recycled construction materials in Europe alone generated revenues of 744.1 million euros in 2010, and this is estimated to reach 1.3 million euros by 2016 [7]. It should be noted that such high figures are conservative as they do not take into account a nearly 100% C&D recycling scenario, which is deemed possible in the near future.

Concrete wastes are one of the common components of the C&D waste produced worldwide. Concrete is the most commonly used construction material, and debris derived from demolished concrete structures constitutes a considerable portion of the C&D waste generated worldwide. Concrete waste has been reported to account for up to 40% of the total C&D waste flux in some countries (Figure 4.1) [8,9].

The huge amount of concrete produced annually causes a number of environmental concerns relating to

- The huge amount of the raw materials extracted and processed to produce the concrete constituents, including aggregates, causing concerns about depletion of nonrenewable natural resources
- The huge amount of energy used at different phases of the concrete life cycle and the associated greenhouse gas emissions
- The huge amount of the waste generated during production and at the end of the concrete service life (demolition wastes)

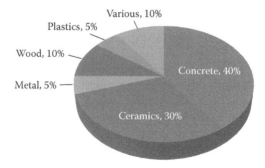

Figure 4.1 Approximate composition of demolition waste in concrete buildings.

Concrete recycling is one of the oldest and most effective waste management strategies for dealing with the enormous amount of concrete debris produced worldwide. Concrete recycling can reduce the costs and energy use incurred in the dumping of debris at remote landfills, reduce landfill space needed, and provide a sustainable source of concrete aggregates by turning concrete debris into suitable aggregates for use in new concrete. The use of recycled concrete aggregates (RCAs) in new construction could reduce considerably the need for extraction of new natural aggregates (NAs). However, the RCAs produced using the current state-of-the-art technology are usually of considerably lower quality than NAs and generally considered not suitable for use in ready-mix concrete for use in structural concrete. The majority of RCAs produced worldwide are currently used as general fills, subbase materials for road construction, and in nonstructural concrete or are mixed in small quantities (up to 20%) with NAs for use in structural concrete. In this chapter, two novel microwave-assisted technologies to improve the quality of RCAs to a level suitable for replacement of NAs for use in structural concrete are introduced. To help readers gain a better understanding of the working principles of these technologies, how they fit into the current concrete-recycling process, and the underlying phenomena leading to RCA quality improvements after microwave processing, this chapter starts by reviewing the current state-of-the-art concrete-recycling technology. Next, the major factors contributing to the lowering of RCA quality compared to NAs are identified. The working principles and efficiency of the microwave-assisted technologies available for improving the quality of RCA through eliminating or alleviating the effects of these factors are then discussed in detail.

4.2 STATE-OF-THE-ART CONCRETE-RECYCLING TECHNOLOGY

The concrete-recycling process starts technically at the "end-of-life" phase of a building's life cycle when the decision to demolish the structure is made. To improve the quality of the recycled concrete products, a soft-strip stage is usually performed prior to demolition. At this stage, all nonstructural components, including mechanical components, sanitary products, glass, wood, and other nonstructural components, as well as the finishes on the concrete floors and walls are removed to minimise the presence of the nonconcrete waste and contaminants present in the concrete debris collected. Next, the building is demolished, and concrete debris is collected for recycling at the recycling facilities.

A concrete-recycling plant refers to a central place where recycling or reprocessing of C&D waste is conducted to obtain the recycled aggregates. The plant generally performs three operations: sorting, crushing, and separating, and may be mobile (on site) or permanent (fixed site). Considering the transportation distance and costs involved, it may make more economic and environmental sense to manage C&D waste locally with a mobile station at the demolition site or at other suitable sites depending on dust, noise, or other such concerns. Whether a mobile or a fixed plant, it is advantageous to locate the site of the recycling operation close to either the sources of raw materials or the end product destinations. The main components of mobile and fixed concrete-recycling plants are briefly described in the following sections.

4.2.1 Mobile recycling plants

Depending on the distance, travelling costs, and volume of concrete debris involved, it may be more economical and environmentally friendly to locate the recycling facilities at the demolition worksite, eliminating the need to transfer the C&D waste to the nearest fixed installation. Mobile C&D waste-recycling plants can make available the traditionally fixed equipment, such as feeders, crushers, magnetic separators, and even vibrating screens, at different locations and at different times. Unlike fixed plants, which are usually connected to the local electricity grid, mobile recycling plants usually rely on diesel generators. This and the lower operational efficiency may in some cases, especially when fixed recycling facilities are available in the vicinity, render mobile plants less environmentally attractive. However, the negative environmental impacts caused by the recycling operations may sometimes be lower for mobile rather than fixed recycling plants when the transportation requirements of the latter and its related environmental impacts, including generation of noise, dust, and gas pollutants typical of motorised vehicular heavy traffic, are taken into consideration. With this in mind, a thorough cost/benefit analysis (economic and environmental) is required to determine the most economical and environmentally friendly recycling strategy for a particular project or demolition worksite.

The components of a typical mobile concrete-recycling plant include feeders, crushers (jaw or impact crushers), built-in conveyor belts, vibrating screens, discharge adjustment systems, and magnetic separators. Commercial mobile concrete-recycling plants are usually available in two forms: track mounted and tyre mounted. The weights and capacities of mobile concrete-recycling plants vary widely, from 14 to 215 tonnes and 50 to 1200 tonnes/h, respectively [10]. A typical mobile recycling plant is shown in Figure 4.2.

4.2.2 Fixed recycling plants

Fixed concrete-recycling plants may be categorised into two general levels based on the quality of the methods deployed for the separation of concrete

Figure 4.2 Mobile concrete recycling plant.

from nonconcrete debris. The simplest plants, which put no or minimal effort into the sorting of debris, are referred to as level 1 fixed plants [11]. This class of concrete-recycling plants achieves the lowest level in terms of recycled output and separation quality. Level 1 plants are similar to mobile plants in terms of the technology used and usually consist of a number of simple crushing steps using either jaw or impact crushers (Figure 4.3), followed by grain size separation. Because of the relatively large amount of

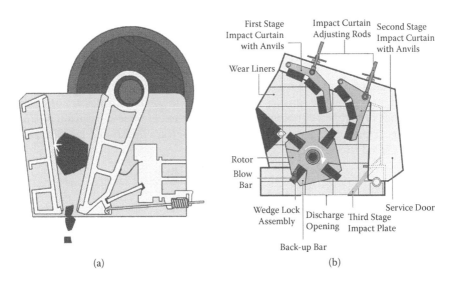

Figure 4.3 Schematic illustration of (a) jaw crusher and (b) impact crusher. (Courtesy of Jalal Afsar, http://www.engineeringintro.com.)

nonconcrete contaminants present, the RCAs produced in level 1 plants are usually usable only for general fills and subbase in secondary roads.

The next class of concrete-recycling plants (level 2) has been developed with the objective of improving the waste separation quality in the plant. In this class of concrete-recycling plants, a variety of separation technologies, including magnets, air sifters, or float separators, are used to separate nonconcrete debris from concrete debris. An air sifter consists of a centrifugal fan and a duct through which the C&D debris is passed. Light materials, including paper, cardboard, plastics, and some wood and nonferrous metals, are blown away and collected as C&D waste mixtures after passing through the duct. Air sifters are known to be ineffective in removing fine and very fine light contaminants. This is because such fine light contaminants are shielded by or attach themselves to the heavier pieces [11]. A float separator is a commonly used alternative separation technique in level 2 plants. In this method, a water-filled bin is used to separate the light contaminants from the heavier concrete debris. Contaminants less dense than water will float and may be easily collected by skimming. Both these separation methods sort all the light nonconcrete contaminants together irrespective of type. Other separation technologies used occasionally in level 2 recycling plants include jiggers and spirals [11]. With the use of separation technologies adopted in level 2 recycling plants, high-purity RCAs are achievable.

Although efficient separation of nonconcrete from concrete debris at the recycling plant is achievable through the use of some or a combination of the separation technologies mentioned, an alternative route to achieve high-purity RCAs is through separation conducted at the source. As explained in Section 4.2, separation of nonconcrete debris from concrete debris before demolition may be performed during the so-called soft-strip stage. While mechanical, electrical, and nonstructural components of buildings can be easily dismantled at this stage, a major difficulty is to achieve the efficient separation of concrete finishes from concrete to which finishes are applied. A number of viable technologies are available to enhance separation during the soft-strip stage. These include a novel microwave-assisted method that is introduced in Section 4.6.1.

4.3 PROPERTIES OF RECYCLED CONCRETE AGGREGATES

To understand how the quality of RCAs and the concrete made with the RCAs, referred to as recycled aggregate concrete (RAC), can be improved using the microwave-assisted technologies introduced further in this chapter, it is essential to be familiar with the basic properties of RCAs produced using the current state-of-the-art technology explained above. In the following sections, some of the results presented in the available literature on the various properties of RCAs, including the adhering mortar content, density, water absorption, and toughness, as well as the effect various production parameters, including the type of the crushing process, particle size, and the strength of the parent concrete, have on these properties are briefly reviewed.

4.3.1 Adhering mortar content

RCAs often contain a large amount of adhering mortar and cement paste. The volume by percentage of old mortar may vary from 20% to 70%, depending on the recycling process, size of RCA particle, and properties of the parent concrete [12,13]. The adhering mortar content is one of the most important properties of RCA because it will directly and indirectly affect all the other major RCA properties [13]. The adhering mortar in RCA may be present in three general forms, and this has been used by some researchers to categorise the RCAs currently produced into three general groups (Figure 4.4). In the first group, referred to as type I RCA, the adhering mortar is present in the form of a layer of cementitious mortar adhering to the surface of a single, usually larger, NA particle. Type II RCA is a conglomerate RCA particle comprising a few, usually smaller, NAs and adhering cementitious mortar. Type III RCA is a conglomerate principally comprised of cementitious mortar. At present, almost all RCAs produced commercially comprise varying proportions of these three types of RCA, and no practical method is currently available to sort the RCA produced efficiently in accordance to types I to III [14]. Adhering cementitious mortar has a lower density, higher water absorption, lower Los Angeles abrasion resistance, and higher sulfate content than NAs used in typical concretes [15,16]. With this in mind, the lower quality of RCAs compared to NAs has been attributed to the presence of cementitious mortar in the RCA produced [15,17,18].

(a)   (b)   (c)

Figure 4.4 Three different types of RCA: (a) type I, comprising one natural aggregate surrounded fully or partially by a layer of mortar; (b) type II, comprising two or more smaller-size natural aggregates surrounded fully or partially by a layer of mortar; and (c) a conglomerate of cementitious mortar.

The negative effects of the presence of mortar on a particular property of RCA are proportional to its content. In addition, the relative severity of the impact of such effects (the proportionality ratio between mortar content and changes in the respective property of RCA) is a function of the ratio between the values of the respective property for the adhering mortar and NA. It has been shown that for RCA particles comprising similar types of embedded NAs and mortar, the relationship between the mortar content and RCA properties is close to linear [15]. The volume of adhering mortar varies with the grain size, strength of the parent concrete, and the crushing process used [12,15]. The effects of these factors are discussed in the following sections.

4.3.1.1 Effects of particle size

The mortar content of RCA varies considerably with the size of RCA particles. In general, the average mortar content of RCA decreases with an increase in the RCA particle size. Fine RCA particles may contain cementitious mortar up to as much as 65% of their total weight, whereas the adhering mortar content of coarse RCA is between 20% and 40% [16,19].

4.3.1.2 Effects of the parent concrete strength

Identifying the effects of the parent concrete's strength on the adhering mortar content of RCA and thereby on the properties of RCA and RAC is still a subject of active research. In 1998, Hasse and Dahms reported that, for the same particle size, RCAs derived from weaker parent concrete have greater dry density and thus less-adhering mortar [20]. Hasse and Dahms explained that this dependency probably stems from the fact that the mortar in weaker parent concrete sheds off more readily during the crushing process, leaving more rock particles that are clean. Contrary to the findings of Hasse and Dahms's study, Grubl and Nealen claimed that the strength of the parent concrete has practically no influence on the strength of the new concrete [21]. Grubl and Nealen asserted that the adhering mortar content of RCA is mainly dependent on the concrete-crushing procedure. They hypothesised that after the first stage of crushing, only the stronger portions of mortar remain adhering.

In our recent study, a series of experiments was designed to investigate the underlying reasons for such discrepancies in the available literature on the effects of the parent concrete's strength on the mortar content of RCA [15]. The results of this study showed that the discrepancies stem mainly from the overlapping effects of other parameters, including the size of the NAs in the parent concrete and the crushing procedure used [15]. It was observed that when the maximum size of NA in the parent concrete was considerably smaller than the maximum size of the RCA particles produced, the mortar content of RCA increased with an increase in the parent concrete compressive strength. This is in agreement with the observations of Hasse and Dahms [20] and may be attributed to the fact that the stronger mortar present in the RCA produced from higher parent concrete strengths results in less mortar being removed during crushing. However, it was also noted that the same relationship does not exist when larger NAs are used in the parent concrete. When the maximum size of NA in the parent concrete was close to the maximum size of the RCAs produced, the effect of mortar content did not seem to follow a consistent increasing trend with an increase in the parent concrete strength.

4.3.1.3 Effect of the crushing process

The crushing process can also significantly affect the properties of the RCAs. The adhering mortar of RCA may be diminished by increasing the number of crushing operations; however, this will increase RCA production costs; hence, a trade-off between the number of crushing stages and RCA quality is required [16].

The effect of the type of crusher has also been studied in the available literature. Jaw crushers (Figure 4.3) reduce large-size pieces by crushing between a set of jaws. A jaw crusher consists of a set of vertical jaws, one fixed and the other moved back and forth relative to it. The jaws are farther apart at the top than at the bottom, forming a tapered chute so that the material is crushed into progressively smaller and smaller sizes as it travels downward until it is small enough to escape through the opening at the base. Impact crushers involve the use of impact forces rather than pressure alone to crush the concrete (Figure 4.3). Concrete is contained within a cage, with openings of the desired size on the bottom, end, or sides that allow the pulverised material to pass through. Impact crushers use a plate hammer mounted on rapidly rotating rotors; high-speed impact forces are generated to crush the concrete pieces in the crushing cavity and propel the crushed concrete pieces along a tangential direction to have an impact on the plate at the other end of the crushing cavity. The concrete pieces are crushed again and are then returned to the plate hammer to undergo the process repeatedly. The concrete pieces bump against each other when propelled between the plate hammer and the impact plate. The concrete pieces become cracked, loosened, and then comminuted by knocking against the plate hammer, impacting with the impact plate, and bumping against adjacent concrete pieces. The crushed concrete with sizes smaller than the gap between the impact plate and the plate hammer will be discharged.

According to Fleischer and Ruby, within the jaw crushers, the concrete chippings are squeezed, thereby inducing incipient cracking and chipping, resulting in a decrease in the strength. They attributed a reduction in frost resistance of the RCA to this action [19]. However, unlike jaw crushers, in impact crushers, the reduction in size takes place without forces being exerted on the individual particles. On the other hand, in impact crushers, the chippings break at their weakest points, such as at incipient cracks, so that the result is a particle with better properties. Reduction by impact also causes the adhering mortar to be removed to a greater extent.

Through a series of comprehensive tests, Akbarnezhad et al. [13] showed that the effects of crushing on the mortar content of RCA depend significantly on the properties of parent concrete, including the strength and size of the NAs used in the parent concrete. Because of the overlapping nature of the mechanisms and interactions of the effects of the various influencing parameters, drawing general conclusions about the relationship between the crushing procedure and RCA properties is difficult. A systematic approach to investigating the effects of one parameter whilst keeping the others constant is necessary to optimise the crushing regime with a view to minimising the mortar content of the RCA produced. By evaluating the properties of the RCAs sourced from two different grades of concrete (30 and 60 MPa) produced with two different coarse aggregate gradings (maximum size of 12 and 20 mm) and crushed using different crushing procedures, Akbarnezhad et al. [13] concluded that, for RCAs sourced from normal-strength concretes (such as C30), the RCA particles produced by crushing the debris to a relatively similar size range as the maximum size of NAs used in parent concrete tended to contain slightly lower amounts of mortar. Therefore, it was recommended that crushing the concrete debris to a maximum size equal to that of their NAs may result in some improvements in the quality of RCA produced. This may be because the smaller difference between the size of the NAs in debris and the space between the jaws of the crusher leads to the production of a smaller amount of RCA particles comprising two or more NAs adhering together and surrounded by mortar (type II) and a higher amount of RCAs comprising a single particle surrounded by a thin layer of mortar (type I), the latter generally having a lower mortar content. However, it was reported that when RCAs are produced from stronger concretes, this trend is not as obvious.

4.3.2 Water absorption and density

Water absorption and density of aggregates are key parameters in the mix proportioning of concrete constituents and are among the simplest and most commonly used tests for determining the overall quality of RCA at concrete-recycling plants and ready-mix concrete plants. Water absorption and density of RCA are interrelated parameters and have been reported to show an almost-linear inverse relationship, mainly because the variations in either of these properties stem from a variation in the mortar content of RCA. As a result of the presence of the intrinsically less-dense mortar, RCAs have been reported to have generally lower bulk density and higher

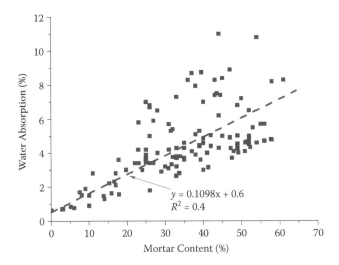

Figure 4.5 Variations in the water absorption of RCA with mortar content.

water absorption than NAs. The variations in the water absorption of RCA with mortar content for a variety of RCA samples reported in available literature are summarised in Figure 4.5.

Available literature suggests that regardless of the properties of the parent concrete, there is a relatively proportional relationship between the mortar content and the bulk density (and water absorption) of the RCAs. The water absorption and bulk density of RCA have been reported to increase and decrease almost linearly, respectively, with an increase in mortar content when considering RCAs produced from similar concrete sources. However, when considering RCAs produced from various concrete sources, the linear proportionality of such relationships do not tend to be as obvious because of the variations in the properties of mortar and the NAs used in the parent concrete.

4.3.3 Toughness (abrasion and impact resistance)

Crushing strength, abrasion resistance, and elastic modulus of aggregate are all interrelated properties that are greatly influenced by porosity [22]. "The abrasion and impact resistance of an aggregate is its ability to resist being worn away by rubbing and friction or shattering upon impact. It is a general measure of aggregate quality and resistance to degradation due to handling, stockpiling, or mixing" [23, p. 17]. The resistance of aggregates against degradation caused by the effects of abrasion, wear, and impact is measured using the Los Angeles test (American Society for Testing and Materials [ASTM] C 131 for aggregates between 2.36 and 37.5 mm [number 8 sieve opening and 1-1/2 in.], and ASTM C 535 for aggregates between 19 and 75 mm [3/4 and 3 in.]). This test subjects the aggregate particles to a combination of impact and abrasion by rubbing against steel balls placed in a slowly revolving drum.

Because of their higher strength and density compared to other components of the concrete, NAs are not normally considered the limiting factor in achieving the overall compressive strength of concrete. However, because of the presence of the adhering mortar, RCAs are generally expected to have lower toughness (Los Angeles abrasion loss > 30%) than NAs. The results of the Los Angeles test conducted on RCAs by Tabsh and Abdelfatah [22] showed that the RCAs had on average 30% more Los Angeles abrasion loss than NAs. Moreover, according to this study, the strength of the parent concrete affects the abrasion resistance of the aggregate (i.e., stronger concrete resulted in less loss). The lower toughness of RCA has also been attributed to the presence of mortar, which typically has a lower toughness compared to NAs. It has been observed that the Los Angeles abrasion resistance of RCA decreases proportionally with an increase in the mortar content [15,16].

De Juan and Gutierrez showed that all samples with adhering mortar content lower than 44% can fulfill the requirements of the Spanish standard for aggregates in concrete. In the Spanish standard, the Los Angeles coefficient has been limited to a maximum of 40% [16]. De Juan and Gutierrez observed that during the Los Angeles abrasion test almost all the adhering mortar of the exposed individual RCAs was pulverised and deposited at the base of the revolving drum.

4.3.4 Impurities

Depending on the sorting technology used at the demolition site or the recycling plant, C&D wastes will constitute varying amounts of nonconcrete impurities, including bricks and tiles, sand and dust, timber, plastics, cardboard and paper, and metals [24]. The effects of contaminants commonly found in C&D wastes (bricks, tiles, clay, glass, wood) on the properties of the RAC has been investigated in the available literature. Poon and Chan showed that the presence of 10% nonconcrete impurities in RCA could result in, on average, a 5% decrease in density of the RAC [25]. For blocks with an aggregate-to-cement (A/C) ratio of 3, the 28-day compressive strengths of contaminated blocks with 100% RCAs were observed to be between 71.5% and 82.3% of the control blocks without any nonconcrete impurities. Dhir et al. showed that mixing 3% plaster with the coarse RCA aggregates led to a 15% reduction in the compressive strength of the concrete produced. They recommended the use of sulphate-resistant Portland cement when recycled aggregates containing plaster or gypsum are used in concrete [17].

4.4 AVAILABLE STANDARDS FOR RCA

With the increasing interest toward the use of recycled aggregates, most major standards bodies have published documentation on the general requirements for the use of RCA and RAC. Among the standards available, the Japanese standard (Japanese Industrial Standards, JIS) is most comprehensive. Japan was among the first countries to initiate research on RCA. Japan introduced recycling laws as early as 1991. Under the "recycled 21" programme launched by the Japanese Ministry of Construction (MOC) in 1992, targets were set to increase the percentage of recycled demolished concrete in the ensuing years. By 2000, demolished concrete was 96% recycled, exceeding the MOC's target of 90%. However, practically all the RCA were used as subbase material for road carriageways.

Previously, the Japanese standard JIS A 5308 for ready-mix concrete did not permit the use of recycled aggregates. In response to the recommendations of the JIS Civil Engineering Committee in 1998, the Japanese Concrete Institute formed a committee to draft a JIS technical report as a preliminary document to be developed into a JIS standard. This report, known as TR A 0006, released in 2000, permits recycled concrete to be used independent of JIS A 5308. However, according to this report, the strength of structural recycled concrete was capped at 18 MPa. Presoaking of the recycled aggregates for the control of workability and the use of blast furnace slag cement or fly ash to counteract alkali aggregate reaction are also included in the requirements of this standard.

Subsequently, in 2005, this report was published as JIS standard JIS A 5021 (Japanese Standard Association, 2005) for high-quality recycled aggregate type "H" for concrete, which is produced through advanced processing, including crushing, grinding, and classifying of concrete masses generated in the demolition of structures. Recycled type H aggregates must have physical properties satisfying the requirements listed in Table 4.1.

Table 4.1 Physical properties requirement for type H recycled aggregates, JIS standard

Items	Coarse aggregate	Fine aggregate
Oven-dry density (g/cm^3)	Not less than 2.5	Not less than 2.5
Water absorption (%)	Not less than 3	Not less than 3
Abrasion (%)	Not more than 35	NA
Solid volume percentage for shape determination	Not less than 55	Not less than 53
Amount of material passing 75-μm test sieve (%)	Not more than 1	Not more than 7
Chloride ion content	Not more than 0.04	

Table 4.2 Limits of amounts of deleterious substances for type H recycled aggregates, JIS standard

Category	Deleterious substances	Limit (mass %)
A	Tile, brick, ceramics, asphalt	2.0
B	Glass	0.5
C	Plaster	0.1
D	Inorganic substances other than plaster	0.5
E	Plastics	0.5
F	Wood, paper, asphalt	0.1
Total		3.0

There are also upper limits set for the amount of deleterious substances present in the type H recycled aggregate, as shown in Table 4.2. JSI A 5023 has also been published as a standard for recycled concrete using low-quality recycled aggregate, type L. This type of concrete produced with type L RCA includes backfilling, filling, and levelling concrete. In addition, the use of type B blended cement is required as a measure against alkali-silica reactivity.

In Hong Kong, two sets of specifications are used for the use of recycled aggregates in concrete production. The specification requirements for RCA are listed in Table 4.3. According to these specifications, 100% recycled coarse aggregate may be used for low-grade applications, and recycled fines are not allowed for use in concrete. The compressive strength for concrete with 100% RAC is capped at 20 MPa, which can be used in benches, planter walls, mass concrete, and other minor concrete structures. For higher-grade applications, up to C35, the Hong Kong specifications allow a maximum of 20% replacement of virgin aggregates by recycled aggregates.

In the United Kingdom, British Standard (BS) 8500-2 and BS EN 12620:2002-A1:2008 have included the general specifications for RCA and RAC in their latest editions. The United Kingdom has not been as active in concrete recycling as some of the other European countries; the older standards, BS 882 and BS 1047, only cover NAs and air-cooled blast slag aggregates, respectively, and did not include any provisions for recycled aggregates. However, these standards have already been withdrawn and replaced with BS EN 12620:2002. The use of recycled aggregates in concrete was first allowed according to BS 8500, published in 2002. According to BS 8500-2, the coarse RCA and coarse recycled aggregate shall conform to the requirements specified in Table 4.4. Composites of coarse RCA or coarse recycled aggregate and NAs shall conform to the general requirements for aggregates as specified in BS EN 12620 and to the general requirements for normal-weight aggregates. Although general recycled aggregates (including nonconcrete waste) can only be used in low-grade concrete not exceeding 20 MPa characteristic strength,

Table 4.3 Specification requirements for recycled concrete aggregate in Hong Kong

Requirements	Limit	Test method
Minimum dry particle density (kg/m^3)	2000	BS 812: Part 2
Maximum water absorption	10%	BS 812: Part 2
Maximum content of wood and other material less dense than water	0.5%	Manual sorting in accordance with BRE Digest 43
Maximum content of other foreign materials (e.g., glass, plastics, clay lumps, asphalt, tar)	1%	
Maximum fines	4%	BS 812: Section 103.1
Maximum content of sand (<4 mm)	5%	BS 812: Section 103.1
Maximum sulfate content	1%	BS 812: Part 118
Flakiness index	40%	BS 812: Section 105.1
10% fines value	100KN	BS 812: Part 111
Grading	Table 3 of BS 882:1992	
Maximum chloride content	Table 7 of BS 8820; 0.5% by mass of chloride ion of combined aggregate	

Note: Data summarizing available standards compiled by Fung, W.K., in The Use of Recycled Concrete in Construction, PhD thesis, University of Hong Kong, 2005.

Table 4.4 Requirements for coarse RCA and coarse recycled aggregate (RA) (mass fraction %), BS 8500-2

Type of aggregate	Requirement[a]					
	Maximum masonry content	Maximum fines	Maximum lightweight material[b]	Maximum asphalt	Maximum other foreign material (e.g., glass, plastics, metals)	Maximum acid-soluble sulfate (SO$_3$)
RCA[a,c]	5	5	0.5	5	1	1
RA	100	3	1	10	1	—[d]

[a] Where the material to be used is obtained by crushing hardened concrete of known composition that has not been in use (e.g., surplus precast units or returned fresh concrete), and not contaminated during storage and processing, the only requirements are those for grading and maximum fines.

[b] Material with a density less than 1000 kg/m^3.

[c] The provisions for coarse RCA may be applied to mixtures of natural coarse aggregates blended with the listed constituents.

[d] The appropriate limit and test method need to be determined on a case-by-case basis.

Table 4.5 Limitations on the use of coarse RCA, BS 8500-2

Type of aggregate	Limitation on use	
	Maximum strength class[a]	Exposure classes[b]
RCA	C40/50	X0, XC1, XC2, XC3, XC4, XF1, DC-1

[a] Material obtained by crushing hardened concrete of known composition that has not been in use and not contaminated during storage and processing may be used in any strength class.

[b] These aggregates may be used in other exposure classes provided it has been demonstrated that the resulting concrete is suitable for the intended environment (e.g., freeze-thaw resisting, sulfate resisting).

concrete containing RCA can have a characteristic strength of up to 50 MPa (Table 4.5). According to this standard, the maximum replacement level of NAs with RCA in designated concretes RC20/25 to RC40/50 should not be more than 20% (by mass) of coarse aggregate except if the specification permits higher proportions to be used. However, for the designed concrete, judgment on the replacement proportion shall be based on the performance required. Furthermore, BS 8500-2 Annex B presents a test method to determine the composition of RCA and recycled aggregate. The use of fine recycled aggregate or RCA is not covered in the British Standard.

In Germany, based on the results of a comprehensive research programme (Baustoffekreislauf im massivbau) conducted between 1996 and 1998, a series of guidelines (Die Ricthlinie) for the use of recycled aggregates in concrete was published. The recommendations of this guideline on the allowable percentage of recycled aggregate for different applications are summarised in Table 4.6. The particle density of the recycled aggregate must

Table 4.6 German guidelines on the maximum percentage of recycled aggregate in relation to the total aggregate

	Recycled aggregates and crushed sand > 2 mm, volume (%)	Crushed sand < 2 mm volume (%)
<B25	35	7
Indoor concrete structures, B35	25	
Concrete for outdoor exposure environment	20	0
Waterproofed concrete		
Concrete with high forest resistance		
Concrete resistant to mild chemical attack		

Source: Fung, W.K., The Use of Recycled Concrete in Construction, PhD thesis, University of Hong Kong, 2005.

not be less than 2000 kg/m³. This guideline is expected to be eventually upgraded to a DIN (German Institute for Standardization) standard [26].

In Austria, the production of recycled aggregate is governed by the *Richtlinie fur recycling-Baustoffe* ("Guidelines for the Recycled Building Materials"), jointly published by the Austrian Association for the Recycling of Building Materials and the Austrian Quality Protection Association for Recycled Building Materials. The third edition of this guideline, published in 1999, covers four types of recycled aggregates: recycled asphalt aggregates; RCAs; mixed recycled asphalt and concrete aggregates; and mixed recycled asphalt, concrete, and NAs. Moreover, three quality classes are defined. Only quality class I RCAs are suitable for use in concrete construction. Aggregates in this class are to be tested for grain size distribution, grain strength, frost resistance, compactability, permeability, foreign material content, contamination, and environmental compatibility as specified in this guideline.

In the Spanish standard, the use of recycled aggregate is restricted to mass concrete and reinforced concrete, and its use for prestressed concrete is prohibited [26]. Moreover, only the use of aggregates from the recycling of conventional concrete is recommended, excluding special types of concrete, such as lightweight concrete, fibre-reinforced concrete, concrete made using aluminous cement, and so on.

European standards are intended to unify practice in the member countries of the European Union, basically to remove trade barriers. BS EN 12620, *Aggregates for Concrete*, was published in January 2003. The merit of BS EN 12620 lies in the way recycled aggregates are treated on an equal basis with NAs. Compliance requirements are implicitly the same for both types of aggregates.

4.5 MAIN FACTORS LOWERING THE QUALITY OF RCA

As discussed, the extensive literature available shows that RCAs produced using state-of-the-art concrete-recycling technology are usually of lower quality compared to NAs and are generally considered unsuitable for use in ready-mix concrete. They are mainly used as base and subbase materials in the construction of secondary roads or mixed in small fractions, up to 20%, with NA to be used in structural concrete [27]. The available literature identifies two factors as the main causes of the lower quality of the RCA compared to NAs:

1. Contaminants (impurities), including chemical and physical contaminants present in the concrete debris or source material used in RCA production
2. The cementitious mortar adhering to the RCA particles, which is of a porous and weak nature

With this in mind, extensive research is ongoing to develop methods for eliminating or reducing the effects of these factors on the quality of RCA. In the following sections, a number of available methods for dealing with these quality-degrading factors are introduced. The focus is placed on two novel microwave-assisted technologies that have been shown to effectively improve the quality of RCA.

4.6 ELIMINATION OF IMPURITIES/CONTAMINANTS

The general term for contaminants used here refers to all materials, hazardous or nonhazardous, that when present in concrete demolition debris, the source material used in producing RCAs, may lower the quality of the RCAs or make them unsuitable for some concrete applications. With this definition, contaminants may include hazardous (chemical or radioactive) contaminants, which may pose a health hazard, as well as nonhazardous contaminants, which may compromise concrete quality (e.g., concrete surface plasters, gypsum, tiles, glazing, etc.). Minimising the presence of such impurities and contaminants in the RCA batches produced may be enhanced through two different routes:

1. Elimination of contaminants at the source (demolition site)
2. Elimination of contaminants at the recycling plant

As discussed previously, elimination at the source is performed during the soft-strip stage prior to the demolition of a building, whereas elimination at the recycling plant relies on postdemolition separation methods to sort and separate the different types of demolition debris delivered at the recycling site. A number of available techniques relating to the second strategy were discussed in Section 4.2. In this section, the focus is placed on the methods available to improve or help in the elimination of the contaminants before building demolition.

Ideally, prior to the demolition of the structure, by incorporating a soft-strip stage, various nonconcrete elements, such as composite roofing, sanitary products, doors, window frames, suspended ceilings, raised floors, carpeting, furnishings, plant and machinery, and so on can be removed in advance before actual demolition on site. However, even if this is the case, because of technical difficulties and the lack of suitable surface removal methods, impurities (such as plasterboard, gypsum, and tiles) present on the surfaces of typical concrete structural elements may not be completely removed efficiently. In addition, depending on the building type and use, the concrete structure may have chemical or physical contaminants (Table 4.7) infused or infiltrated into the structural elements. Such contamination is normally limited to a thin layer on the concrete surface, depending on the age of the structure and length of exposure. Hence, if the contaminated

Table 4.7 Historical risk assessment of building chemical contamination

Historical use	Likely contaminants				
	Inorganic chemicals/Reactions				
	SO	Cl	CO	ASR	pH
Building Type					
Airports	X	✓	✓	X	✓
Carparks	✓	✓	X	X	X
Ceramics works	✓	✓	✓	X	✓
Paints	✓	X	X	✓	✓
Cosmetics	✓	X	X	✓	✓
Disinfectants	✓	✓	X	✓	✓
Explosives	✓	✓	X	X	✓
Fertiliser	✓	✓	✓	✓	✓
Fine chemical	✓	✓	✓	X	—
Sealants	✓	X	✓	✓	✓
Organic chemicals	✓	X	✓	X	✓
Soap manufacturer	X	X	X	X	✓
Dockyards	✓	✓	✓	X	—
Electronics	X	X	X	✓	
Engineering works	✓	✓	✓	X	✓
Gasworks	✓	X	✓	X	✓

Source: From EnviroCentre, Controlled Demolition, National Green Specification. *Demolition: Implementing Best Practice.* Banbury, UK: Waste and Resources Action Programme, 2005.

Note: ASR = Alkali-Silica Reaction.

surface can be efficiently removed, the remaining bulk of the concrete may be confidently used to produce good-quality RCA.

A number of potential mechanical surface removal techniques as well as a novel microwave-assisted selective removal technique were discussed in Chapter 3. These techniques can also be deployed to effectively remove contaminants/impurities from the concrete floors and walls during the soft-strip stage. The selection of the most appropriate method to deploy should be based on the type and extent of contaminant/impurity present. However, although the selected concrete removal techniques described in Chapter 3 can be used on physically or chemically contaminated surfaces during the soft-strip stage, applying such techniques to remove the attached nonhazardous finishes can be costly, both economically and environmentally. This is because the energy required to remove contaminants from surfaces of concrete as part of an integral building is likely to be considerably more compared to when the surfaces are those of broken demolition debris.

With this in mind, much energy and costs can be saved if the bond between the finishes and concrete can be selectively weakened or broken to facilitate removal. A variation of the microwave-assisted selective concrete removal technique (discussed in Chapter 3) appropriately modified to improve surface finish removal efficiency is introduced in the following section.

4.6.1 Microwave-assisted removal of concrete finishes

Ceramic and stone tiles, plasterboard, and gypsum are among the most common types of floor and wall surface finishes used worldwide. Removal of such finishes from concrete floors and walls with a view to reduce nonconcrete waste present in concrete demolition debris is usually carried out through mechanical hacking and other similar demolition techniques. However, such conventional soft-stripping techniques are time consuming and usually accompanied by noise and dust generation, which may be hazardous to the health of occupants and operators. A novel microwave-assisted concrete finish removal technique can considerably improve the efficiency and speed of removal of the concrete surface finishes with reduced health hazards.

Most of the typical concrete finishes, including ceramic and stone tiles, plasterboard, and gypsum and the grout (or mortar) used in levelling or as adhesives have relatively high microwave energy absorption rates and thus are heated rapidly when exposed to microwaves. As discussed previously, the amount of the microwave energy dissipated in dielectric materials, through dielectric loss and the pattern of heating (determined by penetration depth of microwaves), varies significantly with the electromagnetic (EM) properties of the material heated and the microwave frequency. By capitalising on this, we have recently developed a microwave-assisted technique for the removal of concrete surface finishes. The technique makes use of one property of microwaves: The pattern of heating and the penetration depth of the material exposed to microwaves varies with the microwave frequency. By selection of an appropriate range of microwave frequencies, concentrated microwave heating of concrete finishes, including the underlying adhesives or grout, generates a localised field of thermal stresses that is harnessed to separate the finishes and underlying adhesives from floors and walls. To improve the efficiency of this technique, drenching the wall and floor surfaces with water is recommended to maximise energy absorption by the tiles and underlying grout.

The microwave-assisted concrete finish removal equipment we developed is shown schematically in Figure 4.6. The main components of the applicator are shown in Figure 4.7. Some of the main features of the equipment include the following:

1. Portability
2. Built-in remote and local control system

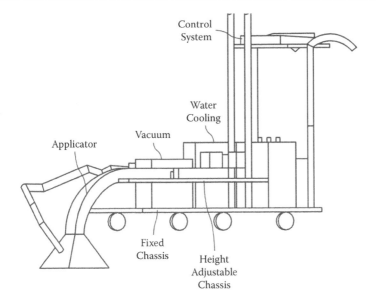

Figure 4.6 Microwave-assisted concrete surface finish removal equipment.

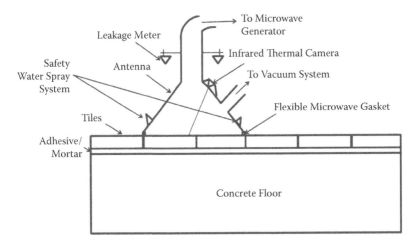

Figure 4.7 Applicator design for typical microwave-assisted concrete surface finish removal equipment.

3. Capability to select the required power and heating duration based on the measured EM properties of the surface finish type
4. Real-time monitoring of the floor temperature using an infrared thermal camera
5. Built-in leakage monitoring and automatic shutdown features that can immediately stop operations in case of excessive leakage

6. Specially designed applicator with attached water spray system to absorb microwaves in the event of excessive leakage and an attached vacuum system to minimise such leakage by reducing the gap between the applicator's gaskets and floor/wall surface by generating a negative pressure within the applicator
7. Frequency range of 2. GHz to 18 GHz
8. Power range of 3 to 10 kW

The recommended operating procedure for removal of ceramic tiles from concrete floors and walls using the microwave-assisted concrete finish removal equipment is summarised in Figure 4.8. As shown, the process starts with the selection of the type of tiles and adhesive from a predefined list (stored in the material library of the control system) of materials with known EM and mechanical properties. Based on the type of material, the control system performs the required simulation analysis to estimate the optimal microwave frequency and power. When the equipment settings are set for optimal operation, the applicator is positioned on the wall or floor surface. A vacuum system attached to the applicator is activated to develop the required negative pressure and thereby ensure a tight surface fit between the applicator gaskets and the floor or wall surface. Microwave heating is then started, and the temperature developed on the concrete surface is continuously monitored using an infrared thermal camera attached to the applicator (Figure 4.7). In addition, as shown in Figure 4.7, a leakage meter attached to the applicator is used to continuously check for microwave leakage to ensure safe operation. If there is undesirable microwave leakage, the control system stops operation and activates the water spray safety system to reduce microwave leakage to allowable limits. Under the International Electrotechnical Commission (IEC) standard, which is applicable to equipment operating in the frequency range from 300 MHz to 300 GHz, power density is measured at least 5 cm from any accessible location on the equipment and should be limited to 5 mW/cm^2 during "normal" operation and 10 mW/cm^2 during abnormal operation.

4.7 REMOVAL OF THE ADHERING MORTAR

As discussed in Sections 4.3 and 4.5, the adhering mortar content is considered the main factor affecting the properties of RCAs and the main reason for the differences in the behaviour of RCA and RAC when compared to NAs and NA concrete, respectively. It is noted that the degree to which the properties of RCAs, including water absorption, density, and toughness, differ from those of NAs is proportional to the adhering mortar content of RCA. With this in mind, lowering the mortar content of RCAs through fine-tuning the production process and separation of mortar from RCAs through

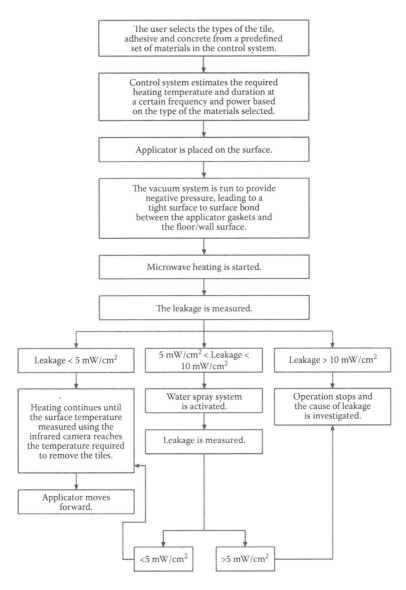

Figure 4.8 A method (framework) for using the mobile microwave-assisted concrete surface finish removal equipment.

postrecycling separation have been studied extensively as two potential routes for improving RCA quality. The studies investigating the first route (i.e., fine-tuning the recycling process) have indicated that the number of crushing stages is crucial in determining the resultant mortar content of the RCA produced [15,16]. It has been shown that the mortar content of RCAs can be reduced considerably by increasing the number of crushing stages. The reduction in the mortar content by incorporating an additional crushing stage varies from 10% to 40% depending on the size of the RCA particles, strength of the mortar, and type of crusher used [15]. However, despite the considerable mortar content reduction achievable, one major disadvantage of incorporating additional crushing stages is the significant decrease in the overall yield of the coarse RCAs produced. This is because, besides the adhering mortar, a considerable portion of weaker embedded original aggregates is also broken into finer aggregate particles. Also, any additional crushing stage introduced adds considerably to the costs of recycling; thus, a trade-off between costs and RCA quality exists.

Alternatively, the second route or strategy for reducing the mortar content of RCAs is postrecycling separation of the mortar present in the RCAs. As the name implies, this form of separation is used to separate the adhering mortar from the RCAs after some or all of the stages of the conventional recycling processes are completed. A number of postrecycling mortar separation methods have been proposed in the available literature. These include thermal separation, mechanical separation, thermal-mechanical separation, chemical separation, and microwave-assisted separation. In the following sections, the working principles as well as the advantages and disadvantages of these methods are discussed briefly. The focus is then placed on the microwave-assisted postrecycling mortar separation method, which has been shown to result in considerable improvements in the properties of RCAs.

4.7.1 Conventional mortar separation techniques

4.7.1.1 Thermal separation

Thermal separation invests on the differences between the thermal expansion rates of mortar and NAs to cause thermal stresses within RCA. In thermal separation, RCA particles are heated to temperatures ranging from 300°C to 600°C, depending on the strength of mortar and NA type, for about 2 hours to break up and separate the mortar [28,29]. Mortar in RCA typically has a higher thermal expansion rate than NAs and thus expands faster than NA when heated. Therefore, considerably higher thermal stresses develop in the mortar than in the NA when RCA is heated. Moreover, the differences in the expansion rate of NA and mortar is expected to lead to considerable differential thermal stresses at their interface. These mechanisms together with the intrinsically weaker nature of typical mortar compared to typical NA are mobilised to break the adhering mortar into fine powder, thereby separating mortar present in the RCA. A number of studies have recommended that presoaking the RCA in water to saturate the adhering mortar can result in higher separation efficiency. This is deemed to be because of higher internal pore water pressures developing

Table 4.8 Efficiency of the thermal separation method

		Properties of RCA		
Type of RCA	Process duration (h)	24-h water absorption (%)	Bulk density (OD) (kg/m³)	Mortar content (%) by mass
Before separation	0	4.2	2370	47
After thermal separation 300°C	2	4.1	2380	44
500°C	2	3.8	2390	41
600°C	2	3.8	2390	38

Note: OD = oven-dried.

in the mortar as a result of rapid evaporation of the internal pore water when RCA is exposed to temperatures above 100°C. In addition, the rapid cooling of RCAs after heating by immersion in cold water (cold relative to the RCA temperature) has been reported as another effective method to improve the efficiency of thermal separation by increasing the differential thermal stresses developed [16].

The efficiency of such thermal separation techniques has been investigated and reported in a number of previous studies [30,31]. A study we conducted (Table 4.8) showed that heating the RCA at 300°C, 500°C, and 600°C for 2 hours reduced the mortar content by 6.4% (from 47% to 44%), 12.8% (from 47% to 41%), and 19.1% (from 47% to 38%), respectively.

Thermal separation of mortar is one of oldest and most common postrecycling separation strategies for reducing the mortar content of RCA. However, it has significant disadvantages with regard to efficiency as well as environmental and economic impacts. As a result of the long periods of heating at high temperatures, thermal separation of mortar requires relatively high energy consumption with the associated undesirable environmental emissions. Such negative environmental impacts tend to reduce the perceived environmental advantages of concrete recycling. In addition, as a result of the uniform heating of RCA in the thermal separation method, relatively high thermal stresses are expected to be developed in the original NA present in RCA, which could damage the stone aggregate itself. For instance, when the NA in RCA is granite, heating the RCA at 400°C and 600°C can result in about 14% and 44% reduction in the pressure resistance of the granite aggregate, respectively [32]. Therefore, the removal of mortar from RCA through thermal separation is recommended only when NA present in RCA is substantially stronger than the adhering mortar.

4.7.1.2 Mechanical separation

Mechanical separation methods rely on mechanical forces in the form of rubbing and impact to separate the adhering mortar. In a method known

Table 4.9 Efficiency of the mechanical separation (mechanical rubbing) method

Type of RCA		Process duration (h)	24-h water absorption (%)	Bulk density (OD) (kg/m³)	Mortar content (%) by mass
			Properties of RCA		
Before separation		0	4.2	2370	47
After mechanical separation	100 revolutions	~0.1	3.5	2410	34
	200 revolutions	~0.2	3.0	2420	28
	300 revolutions	~0.3	2.8	2440	17

as eccentric-shaft rotor, RCAs are passed through rubbing equipment comprising an inner and outer cylinder that rotate eccentrically at a high speed. The RCAs are rubbed against the cylinder walls and each other, breaking the adhering mortar into a fine powder, which is collected after passing through the 2- to 4-mm sieves provided on the surface of the inner cylinder. In another mechanical separation method referred to as the grinding method, a drum containing iron balls is used to provide the required impact and the friction forces required to separate mortar from RCA. In this method, RCAs are rubbed against each other and the iron balls housed in the rotating portioned sections [33].

The results of a study we conducted showed that 100, 200, or 300 revolutions of a mechanical rubbing drum equipment filled with 10 steel balls resulted in about 28%, 40%, and 64% reduction in the mortar content of RCA, respectively (Table 4.9). As can be seen, relatively high mortar removal rates can be achieved using the mechanical separation method, and the removal rate tends to increase with an increase in the number of iron balls and drum revolutions. The rate, however, is achieved at the expense of higher energy consumption and costs; therefore, a trade-off has to be made between the costs incurred in adopting particular regimes of the separation process and the quality of RCA produced at the end of the process. Mechanical separation is easy to use and more efficient in the removal of mortar compared to the thermal separation method. However, it comes with environmental impacts arising from the relatively high energy consumption required and inherently high noise pollution. This method also has a relatively lower yield of coarse aggregates compared to other methods because mechanical rubbing and impact also tend to break a considerable portion of the coarse NA particles into finer particles.

4.7.1.3 Thermal-mechanical separation

The thermal-mechanical class of methods uses a combination of thermal stresses generated through conventional heating at high temperatures and

Table 4.10 Efficiency of the thermal-mechanical separation method

Type of RCA		Process duration (h)	Properties of RCA		
			24-h water absorption (%)	Bulk density (OD) (kg/m³)	Mortar content (%) by mass
Before separation		0	4.2	2370	47
After thermal-	300°C	~2.1	3.3	2430	31
mechanical	500°C	~2.1	2.1	2480	21
separation					

the mechanical stresses generated through the application of mechanical separation methods to separate the adhering mortar from RCA [28,29]. In a typical combined thermal-mechanical separation method, referred to as the heating-and-rubbing method, the adhering cementitious mortar is dehydrated and thus becomes brittle by heating in a vertical furnace. The heated RCA is then transferred to rubbing equipment, which comprises a tube-type mill with inner and outer cylinders loaded with a number of iron balls to remove the adhering mortar. Shima et al. (2005) showed that the heating-and-rubbing method may increase the quality of RCA significantly, so it could comply with the quality requirements of the Japanese Concrete Institute (JCI) for the high-quality H RCA [28]. The results of a study we conducted (Table 4.10) show that the heating of RCAs in a conventional oven at 500°C for 2 hours followed by mechanical rubbing using a Los Angeles testing machine loaded with 10 iron balls for 100 revolutions could result in 55% reduction in the mortar content of RCA.

However, despite the report of such promising separation rates, this combined method tends to be high in energy consumption and costly as well. The negative impacts in terms of energy and associated carbon emissions could easily negate the perceived environmental advantages of concrete recycling. The separation efficiency of the combined thermal-mechanical method can also be increased by increasing the duration and temperature of heating, the number of iron balls, and the number of revolutions of the drum but at the expense of additional costs and energy consumption. A detailed economic and environmental analysis is required before this method can be adopted optimally.

4.7.1.4 Chemical separation

Because of the alkaline nature of cement, cementitious mortar can be easily corroded by strong acids [34]. The chemical separation method uses the inherently weak corrosion resistance of the cementitious mortar to separate the adhering mortar from NAs in RCA. In this method, RCAs are immersed in diluted acids for about 24 hours and then washed to remove

the corroded mortar. The selection of an appropriate acid is highly important and could significantly affect the effectiveness of the chemical separation method. Sulphuric (H_2SO_4) and hydrochloric (HCl) acids have been
reported as the most efficient acids for the removal of mortar from RCA [35].

However, besides the mortar removal efficiency, another important
parameter in choosing a suitable acid for chemical separation methods is
the compatibility of acid with the NAs present in the RCA. The acid used
should have no or minimal effect on the quality of the natural stone aggregates present in RCA. For instance, for RCAs with granitic embedded NAs,
HCl and H_2SO_4 are considered the most appropriate choices because of the
considerably low solubility of the component minerals present in granitic
aggregates when exposed to these acids [36,37]. On the other hand, hydro-
fluoric acid should not be used as all the major constituents of industrial
granites, including quartz, feldspar, and mica, can be easily dissolved in it
[36,37]. In general, the chemical separation method has been suggested to
be especially suitable for RCAs consisting of NAs with high chemical resistivity, such as granite [14,36].

The efficiency of the chemical separation method has been reported to
depend on the porosities of the adhering mortar and embedded NAs, acid
type and concentration, ratio between the volumes of acid and RCAs being
processed, temperature, process duration, and the container type (static vs.
dynamic). The results of a study we conducted (Tables 4.11 and 4.12) show
that, for a similar acid (sulphuric acid) and similar process duration, the
removal rate increases significantly with an increase in the acid concentration or the volumetric ratio between the acid and RCA, both leading to an
increased presence of the H^+ ions required for corrosion of the RCA to take
place [13]. However, it was observed that there was a particular H^+ concentration associated with each test (different concentrations and different
acid/RCA volumetric ratios) after which the beneficiation effects of further

Table 4.11 Different chemical separation methods investigated

	Chemical separation methods			
Action taken	I	II	III	IV
Soaking in sulfuric acid solution (for a total duration of 24 h)	×	×	×	×
Continuous rotary agitation of RCA/acid container (10 ± 1 rpm)			×	×
Washing away the corroded mortar after 8 h		×		×
Replacement of the acidic solution with a fresh acidic solution (after 8 h)		×		×
Washing and cleaning of RCA samples on 4-mm sieve after 24 h	×	×	×	×

Note: Results are summarized in Table 4.12.

Table 4.12 Efficiency of the chemical separation methods described in Table 4.11

Acid ([H⁺]) concentration (mol/L)	Mortar removed/Total mortar (%, by mass)							
	Method I Vacid/VRCA		Method II Vacid/VRCA		Method III Vacid/VRCA		Method IV Vacid/VRCA	
	2.5	5	2.5	5	2.5	5	2.5	5
1	12 ± 4	22 ± 6	27 ± 5	45 ± 8	23 ± 8	79 ± 6	37 ± 4	86 ± 7
2	23 ± 9	54 ± 8	33 ± 4	79 ± 6	35 ± 5	86 ± 5	44 ± 5	94 ± 4
3	35 ± 8	73 ± 4	57 ± 7	88 ± 2	47 ± 10	85 ± 8	42 ± 7	~100
4	43 ± 8	68 ± 2	59 ± 5	85 ± 8	44 ± 7	91 ± 4	67 ± 7	~100
5	51 ± 4	70 ± 7	55 ± 4	91 ± 4	56 ± 3	82 ± 6	77 ± 9	~100
6	48 ± 5	75 ± 7	50 ± 8	88 ± 3	55 ± 5	89 ± 3	70 ± 3	~100

increases in acid concentration were reduced considerably. This is probably because once a sufficient amount of H⁺ ions is present, acid corrosion of the RCA is mainly controlled by the permeability of the mortar. The permeability of the adhering mortar tends to decrease gradually with successive stages of acid exposure as the silica and aluminosilicate gels released by C-S-H cover the exposed surfaces of the RCA particles [36]. As shown in Table 4.12, the efficiency of the chemical separation method may also be increased considerably through the use of a suitable rotary agitation system or addition of a washing stage to remove the previously corroded mortar from the surface of the RCA. This is because these methods could considerably increase the accessibility of the acid to reach additional unexposed mortar as fresh surfaces are then exposed for further corrosion of the RCA to take place [13].

Chemical separation using strong acids at high concentrations is generally considered an efficient technique for use in separating adhering mortar from RCA. However, one major problem preventing the widespread use of this technique in practice is the potential detrimental effects any residual acids remaining on the RCA have on the durability of concrete. Traces of residual sulphuric and chloride acids, if present, can considerably increase the sulphate and chloride contents of the RCA and thus lead to a decrease in durability of the RAC. Such durability concerns could be addressed using low-concentration acids (~0.1 molar), but this is at the expense of efficiency [35]. Table 4.12 shows that the removal mortar rate achievable at low acid concentrations is rather marginal. In addition, the relatively long processing time required (>24 hours) is another major disadvantage of the chemical separation method when compared to other separation methods discussed. As a result of the significant durability concerns and long processing time required, chemical separation at high acid concentrations has been proposed mainly as an effective method to achieve complete removal of mortar from RCA particles for the accurate measurement of the RCA mortar content in laboratory test samples rather than as a bona fide separation technique in itself [13].

4.7.2 Microwave-assisted separation

Different separation methods based on conventional heating, mechanical rubbing, and acid treatment or a combination of these methods were discussed in the previous sections. As discussed previously, a more efficient alternative to these methods is the microwave-assisted separation method. In the following, we start by describing the working principles of the microwave-assisted mortar separation method and reasons behind its exceptional efficiency in improving the quality of RCA. Next, the available literature on the improvements achievable using this novel separation technique and its efficiency compared to other separation methods discussed previously is reviewed.

4.7.2.1 Working principles

An RCA particle typically comprises two main components, the embedded NAs and adhering cementitious mortar, which are both dielectric materials and therefore are heated because of dielectric losses when exposed to microwave power. As discussed, the extent to which a dielectric material is heated when exposed to microwaves and the heating pattern depend on microwave frequency, microwave power, and most important, the EM properties of the material. An important property commonly used to estimate the heating rate of dielectric materials in a microwave field is the attenuation factor β. In general, in typical microwave heating cases, the heating rate increases exponentially with the attenuation factor. As shown in Figure 4.9, typically

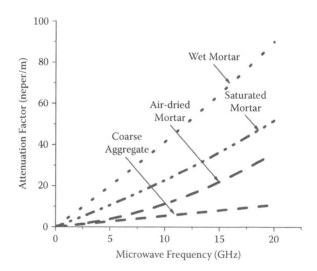

Figure 4.9 The attenuation factor of mortar and natural aggregate.

mortar has a higher attenuation factor and is therefore heated faster than natural stone aggregates when placed in a microwave field.

Another important point to be taken into account is the considerable increase in the attenuation factor of mortar with an increase in its water content. As a result, an increase in the water content of mortar could considerably increase the differences between the attenuation factor and thus the heating rate of the mortar and natural stone aggregates. On the other hand, mortar is of a more porous nature and thus has higher water absorption than NAs. Therefore, saturating the adhering mortar through soaking the RCA particles in water for a few minutes can be used to increase the differences in the water content and thus the differences between the microwave heating rate of the adhering mortar and NAs. In the microwave-assisted separation method, these inherent differences between the dielectric properties and water absorption rates of the mortar and NAs are used to generate a localised field of differential thermal stresses in the mortar, especially at the interface between mortar and NA present in the RCA in a relatively short duration (a few seconds to a few minutes, depending on the microwave power and the volume of the RCA processed) without causing a significant temperature rise in the NA itself [31,38].

To elaborate, let us revisit the microwave power dissipation formulations discussed in Chapter 1. The microwave power dissipation in a dielectric material may be estimated using Lambert's law. The simple form of Lambert's law may be stated as

$$PL(x) = -\frac{\partial I(x)}{\partial x} = 2\beta I_0 e^{-2\beta x} \tag{4.1}$$

Considering an RCA particle exposed to microwaves, here $PL(x)$ is the microwave energy dissipated at a distance x from the microwave-exposed surface of RCA, I_0 is the microwave power transmitted into RCA, and β is the attenuation factor of RCA. As can be seen in Equation 4.1, a higher attenuation factor results in higher microwave power dissipation and thus faster decay of the microwave energy in the material. Consider a material comprising two layers with two different attenuation factors, β_1 and β_2, and two different thicknesses, L_1 and L_2, respectively, exposed to uniform microwave power from its top surface. The microwave power dissipation in the upper layer (layer 1) may be calculated as

$$PL_1(x) = 2\beta_1 I_0 e^{-2\beta_1 x} \tag{4.2}$$

The dissipated microwave power at distance x from the top surface of the next layer (layer 2) is calculated through

$$PL_2(x) = 2\beta_2 \left(I_0 - \int_0^{L_1} PL_1(x) \right) e^{-2\beta_2 x} \qquad (4.3)$$

Now, consider that layer 1 has a considerably higher attenuation factor compared to layer 2. In this case, according to the equations, more energy tends to be dissipated in layer 1; thus, layer 1 would be heated much faster than layer 2.

$$\beta_1 > \beta_2 \qquad (4.4)$$

and

$$I_0 > I_0 - \int_0^{L_1} PL_1(x) \rightarrow 2\beta_1 I_0 \gg 2\beta_2 \left(I_0 - \int_0^{L_1} PL_1(x) \right) \rightarrow PL_1(x) \gg PL_2(x)$$

Hence, in this case, microwave heating leads to differential heating, especially at the interface of the two layers. Now, consider RCA as a composite material comprising two layers: the adhering mortar and the recycled NA. For a similar moisture condition, the attenuation factor and thus the microwave energy absorption rate of mortar is higher than for NAs (Figure 4.9). In addition, as mentioned, the difference in the microwave absorption rate of NA and mortar could be considerably increased by increasing the water content of mortar. Hence, if RCA is exposed to microwaves, the mortar layer would be heated much faster than the NA; thus, significant differential thermal stresses may develop at the mortar-aggregate interface.

In the microwave-assisted separation method, these differential thermal stresses may be harnessed to remove the adhering cementitious mortar from RCA and thereby increase the quality and yield of the RCA products. Moreover, it is well known that the water-to-cement (w/c) ratio increases at the interfacial transition zone (ITZ) and can be significant if bleeding is predominant; hence, the ITZ normally has higher porosity and water absorption than the bulk cementitious mortar. As a result of the higher water content, the ITZ is expected to heat up even faster when RCA is exposed to microwaves, leading to higher differential thermal stresses at the ITZ. Besides the differences between the water absorption and the EM properties of mortar and aggregate, the differences in the thermal properties, such as the specific heat, thermal conductivity, and expansion coefficient, may also contribute to the generation of differential stresses at the mortar-aggregate interface when RCA is subjected to microwave heating. However, the effects of these properties are considered to be much less significant compared to the effects of the different EM properties [31].

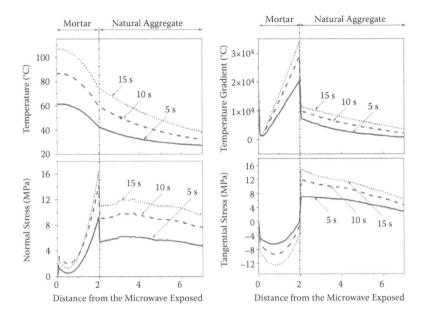

Figure 4.10 Temperature, temperature gradient, normal stress, and tangential stress developed in RCAs processed with microwave-assisted mortar separation equipment. (From Akbarnezhad, A., Ong, K.C.G. et al., *Construction and Building Materials*, 2011, **25**:3469–3479. With permission from Elsevier.)

Figure 4.10 shows the results of a numerical study we conducted to illustrate the underlying phenomena leading to the separation of adhering mortar from RCA in the microwave-assisted separation method [31]. The RCA considered in this study was a 14-mm diameter, perfectly spherical particle; it comprised a 2-mm thick layer of adhering mortar covering a 10-mm diameter sphere of natural granite aggregate. The RCAs were assumed to be exposed to a constant microwave power of 10 kW at 2.45-GHz microwave frequency. The configuration of the microwave-assisted mortar separation equipment considered in the simulation is shown in Figure 4.11.

As shown in Figure 4.10, with only 5 to 15 seconds of microwave heating using a typical industrial microwave oven operating at 10-kW power, considerably high tensile stresses in the range of 10 to 16 MPa can be developed at the mortar-NA interface, which can be harnessed to separate the adhering mortar from the NAs. As shown, because of the considerably lower heating rate of NA compared to mortar, the tensile stresses developed in the NAs are likely to be lower than those developed in the mortar. Therefore, by considering the expected higher strength of the typical aggregates compared to mortar, any damage to the integrity of NAs at the time of mortar fracture or detachment seems unlikely unless inherent fissures are already

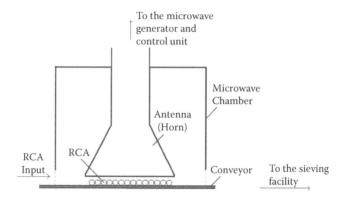

Figure 4.11 Configuration of the RCA mortar separation equipment considered in the numerical study.

present. Of course, to ensure minimal destructive effects of the microwave-assisted separation method on NAs in RCA, the operation should be sufficiently fine-tuned to avoid excessive heating in terms of temperature and time duration.

The effectiveness of the microwave separation method has also been validated by a number of experimental studies. The results of a study we conducted are presented in Table 4.13. In this experiment, 2-kg samples of RCA were heated at maximum power for 1 minute using a 10-kW industrial microwave oven. The microwave-processing equipment used is shown in Figure 4.12. The heated samples were then immediately quenched in 25°C water. Rapid cooling of heated samples could also improve the mortar separation rate by generating additional differential thermal shock stresses. To investigate the effects of RCA water content, samples were tested at two different water contents; saturated or air dried. A sample of RCA before and after microwave processing is shown in Figure 4.13.

It was observed that microwave heating of saturated RCA samples at 10 kW for a duration of 1 minute resulted in almost 48% reduction in the mortar content. The decreased mortar content consequently led to an almost 33% decrease in the water absorption and 3.8% increase in the bulk

Table 4.13 Efficiency of the microwave-assisted mortar separation technique

Type of RCA		Process duration (h)	24-h water absorption (%)	Bulk density (OD) (kg/m³)	Mortar content (%) by mass
			Properties of RCA		
Before separation		0	4.2	2370	47
After microwave separation	Presaturated RCA	~0.02	2.8	2460	24
	Air-dried RCA	~0.02	3.4	2430	32

(a)   (b)

Figure 4.12 Microwave-assisted mortar separation equipment.

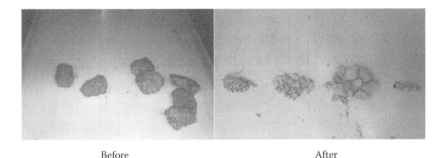

Before   After

Figure 4.13 RCAs before and after microwave separation.

density of the RCAs treated. Microwave heating of air-dried RCA particles led to about 32% reduction in the mortar content and thereby 19% reduction in the water absorption and 2.5% increase in the bulk density of RCAs [31]. Additional rounds of microwave processing or an increase in the microwave power and microwave heating duration can result in more complete removal of the adhering mortar from RCA; however, it should be noted that this may compromise the perceived economic and environmental benefits derived; therefore, the optimal level of microwave heating should be assessed prior to large-scale implementation.

The infrared thermal image of the RCAs taken immediately after the microwave separation process is shown in Figure 4.14. As shown, the maximum temperature at the time of mortar fracture or detachment was less than

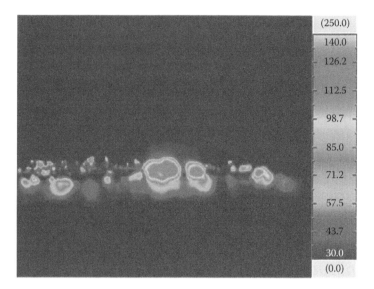

Figure 4.14 Temperature of RCAs and separated mortar pieces after the microwave separation process.

150°C, which is unlikely to affect the properties of the natural stone aggregates after the microwave separation process. This may be considered one of the major advantages of microwave-assisted mortar separation over the thermal separation method. Another advantage of the microwave-assisted mortar separation method is its short processing time, which results in considerably lower energy consumption and associated emissions compared to thermal, mechanical, and thermal-mechanical separation methods. In addition, unlike the chemical separation method, the microwave-assisted mortar separation method poses no additional concrete durability risk. These advantages render the microwave-assisted mortar separation method an appealing technique in producing high-quality RCAs.

4.7.2.2 Properties of RACs made with microwave-treated RCAs

Section 4.2 noted that the basic properties of RCA are directly related to its mortar content. Therefore, the reduction of mortar content is expected to proportionally improve the quality of RCAs produced. Consequently, any reduction in the mortar content of RCA is also expected to lead to an improvement in RAC performance. The results of an experimental study we conducted that was aimed at investigating the effects of incorporating various replacement percentages of microwave-treated RCA for coarse NA are shown in Figure 4.15 [31]. As shown, the results of this study indicate

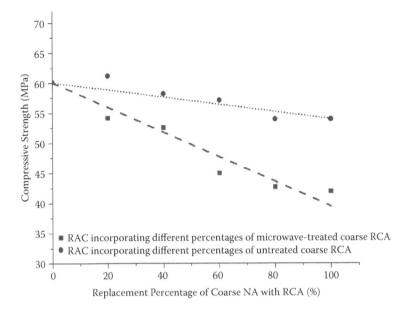

Figure 4.15 Variation in the compressive strength of RAC with an increase in its RCA content. (From Akbarnezhad, A., Ong, K.C.G. et al., *Construction and Building Materials*, 2011, **25**:3469–3479. With permission from Elsevier.)

a generally inverse relationship between the compressive strength of recycled concrete and the replacement percentage of RCAs (both untreated and microwave-treated RCAs) for the coarse aggregates. As can be seen, the reduction in the mortar content of RCAs after microwave separation led to considerable improvements in the properties of RACs. Incorporating 100% coarse microwave-treated RCA in place of NAs led to only a 10% reduction in the compressive strength of concrete as compared to almost 30% reduction when a similar amount of untreated RCA was used. Moreover, the effects of incorporating up to 40% microwave-treated RCA on the compressive strength of RAC seemed to be negligible.

Similarly, we observed that a reduction in mortar content of RCA through microwave separation could result in considerable improvements in other mechanical properties of RAC, including the modulus of elasticity and flexural strength. The modulus of elasticity and flexural strength of concrete were reduced by only about 10% when all NAs were replaced with microwave-treated RCA. This is considerably better in terms of concrete performance when compared with the up to 25% and 15% decrease in the modulus of elasticity and flexural strength, respectively, when untreated RCAs were incorporated. It has been debated that besides the decrease in the mortar content, the positive effects of microwave processing of RCAs are also partially attributable to the breaking of the weaker RCA particles as a result of the stresses developed during the microwave heating process.

As illustrated, the microwave mortar separation method introduced in this chapter can serve as an efficient tool for improving the quality of RCA to levels acceptable for use in structural concrete. However, it should be noted that, like other separation techniques described previously, the microwave separation method also comprises a number of stages during processing that may result in an increase in the overall costs, energy use, and carbon footprint when adopted for recycling RCA. With this in mind, it is important to evaluate the economic and environmental impacts of the additional separation processes for optimisation before full implementation of the microwave-assisted mortar separation method. One example is the soaking process. Too much water would lead to energy wastage in unnecessarily turning excess soak water to steam. Too little would render the whole process less than efficient.

4.8 SUMMARY

Concrete recycling is an increasingly common method of disposing of demolition rubble and can provide a sustainable source of concrete aggregates by crushing concrete debris into specific size fractions that can be used as aggregates in concrete. It has been shown that up to 90% of structural concrete debris may be used to produce recycled aggregates of an acceptable quality. Besides serving as an alternative source of aggregate, concrete recycling may also reduce the landfill space needed to dispose of construction debris. In addition, lower transportation cost and reduced environmental impact are among other advantages of concrete recycling. The use of RCAs in construction works is a subject of high priority in the building industry throughout the world. However, the RCAs currently produced are usually of lower quality compared to NAs and are therefore not suitable for use in ready-mix concrete. They are mainly used as base and subbase materials in road carriageway construction or mixed in small fractions, up to 20%, with NAs to be used in ready-mix concrete. The nonconcrete impurities and adhering mortar present in RCA have been identified as main causes of lower-quality RCA compared to NAs.

In this chapter, a number of methods available to increase the quality of RCA through eliminating or reducing the negative effects caused by these parameters were reviewed. In addition, two novel microwave-assisted techniques capable of achieving better RCA quality and eliminating the drawbacks of existing beneficiation methods were introduced. In the first technique, a rather similar concept as that used in microwave-assisted concrete demolition is used to remove the concrete surface finishes before demolition of a building. This removal stage can significantly reduce the amount of nonconcrete impurities present in the concrete debris. The second technique, termed microwave-assisted mortar separation, takes advantage of the inherently higher dielectric loss of mortar than typical stone aggregates to selectively heat and thereby remove the mortar portion of RCA. The fundamentals and working principles of these microwave-assisted methods used in concrete recycling were discussed in detail. Furthermore, the effects of microwave-assisted mortar separation on the properties of RCA and RAC were discussed.

REFERENCES

1. Anik, D., Boonstra, C., and Mak, C., *Handbook of Sustainable Building: An Environmental Preference Method for Selection of Materials for Use in Construction and Refurbishment.* London: James & James, 1996.
2. Kulatunga, U. and Aaratunga, D., Attitudes and pereptions of construction workforce on construction wastein Sri Lanka. *Management of Environmental Quality*, 2006, **17**(1):57–72.
3. Yuan, H.P., Chini, A.R., et al., A dynamic model for assessing the effects of management strategies on the reduction of construction and demolition waste. *Waste Management*, 2012, **32**(3):521–531.
4. Sonigo, P., Hestin, M., and Mimid, S., Management of construction and demolition waste in Europe. Stakeholder Workshop, Brussels, 16 February 2010.
5. McDonald, B. RECON waste minimization and environmental program. In *Proceedings of CIB Commission Meetings and Presentations.* Melbourne, Australia, RMIT, 1996, 14–16.
6. United States Environmental Protection Agency (EPA), *Resource Conservation Challenge: Campaigning Against Waste*, EPA 530-F-02-33. Washington, DC: EPA, Solid Waste and Emergency Response, 2002.
7. Forst and Sullivan. Strategic Analysis of the European Recycled Materials and Chemicals Market. 2011. http://www.forst.com/prod/servlet/report-toc.pag?repid-570-01-00-00-00 (accessed December 9, 2013).
8. Coelho, A. and Brito, J., *Analysis of the Viability of Construction and Demolition Waste Recycling Plants in Portugal—Part III: Analysis of the Viability of a Recycling Plant.* 2012, ICIST DTC 12/12 report. Lisbon, Portugal: ICIST, Instituto de Engenharia de Estruturas, Território e Construção (Institute of Structural Engineering , Territory and Construction).
9. Oikonomou, N.D., Recycled concrete aggregates. *Cement and Concrete Composites*, 2005, **27**(2):315–318.
10. Metso Minerals, *Crushing and Screening Handbook.* 4th ed. Helsinki, Finland: Metso Minerals, 2009.
11. Coelho, A. and De Brito, J., Preparation of concrete aggregates from construction and demolition waste (CDW), in *Handbook of Recycled Concrete and Demolition Waste*, Pacheco-Torgal, F., Tam, V., et al., eds. Cambridge, UK: Woodhead, 2013, 181–185.

12. Li, X., Recycling and reuse of waste concrete in China. Part I. Material behaviour of recycled aggregate concrete. *Resources, Conservation and Recycling*, 2008, **53**(1–2):36–44.

13. Akbarnezhad, A., Ong, K.C.G., et al., Acid treatment technique for determining the mortar content of recycled concrete aggregates. *Journal of Testing and Evaluation*, 2013, **41**(3):441–450.

14. Akbarnezhad, A., Ong, K.C.G., et al., Beneficiation of recycled concrete aggregates. *Concretus (SCI)*, 2010, **2**(1):14–17.

15. Akbarnezhad, A., Ong, K.C.G., et al., Effects of the parent concrete properties and crushing procedure on the properties of coarse recycled concrete aggregates. *Journal of Materials in Civil Engineering*, 2013, **25**(12):1795–1802.

16. De Juan, M.S. and Gutierrez, P.A., Study on the influence of attached mortar content on the properties of recycled concrete aggregate. *Construction and Building Materials*, 2009, **23**(2):872–877.

17. Dhir, R., Paine, K., and Dyer, T., Recycling construction and demolition wastes in concrete. *Concrete (London)*, 2004, **38**(3):25–28.

18. Dos Santos, J.R., Branco, F., and De Brito, J., Mechanical properties of concrete with coarse recycled aggregates. *Structural Engineering International: Journal of the International Association for Bridge and Structural Engineering (IABSE)*, 2004, **14**(3):213–215.

19. Fleischer, W. and Ruby, M., Recycled aggregates from old concrete highway pavements. In *Proceedings of International Symposium-Sustainable Construction: Use of Recycled Concrete Aggregate*, London, November, 1998.

20. Haase, R. and Dahms, J., Material cycles on the example of concrete in the northern parts of Germany. *Beton*, 1998, **48**(6):350–355.

21. Grubl, P. and Nealen, A. Construction of an office building using concrete made from recycled demolition material. In *Proceedings of International Symposium-Sustainable Construction: Use of Recycled Concrete Aggregate*, London, November, 1998.

22. Tabsh, S.W. and Abdelfatah, A.S., Influence of recycled concrete aggregates on strength properties of concrete. *Construction and Building Materials*, 2009, **23**(2):1163–1167.

23. Concrete Knowledge Center, *Aggregates for Concrete*. Farmington Hills, MI: American Concrete Institute, 2007.

24. Shicong, K., Reusing Recycled Aggregates in Structural Concrete, PhD thesis, Hong Kong Polytechnic, Hong Kong, 2006, 312.

25. Poon, C.-S. and Chan, D., Effects of contaminants on the properties of concrete paving blocks prepared with recycled concrete aggregates. *Construction and Building Materials*, 2007, **21**(1):164–175.

26. Fung, W.K., The Use of Recycled Concrete in Construction, PhD thesis, University of Hong Kong, 2005.

27. Shayan, A. and Xu, A., Performance and properties of structural concrete made with recycled concrete aggregate. *ACI Materials Journal*, 2003, **100**(5):371–380.

28. Hirokazu, S., Hisashi, T., et al., An advanced concrete recycling technology and its applicability assessment through input-output analysis. *Journal of Advanced Concrete Technology*, 2005, **3**(1):53–67.

29. Noguchi, T., Toward sustainable resource recycling in concrete society. In *Second International Conference on Sustainable Construction Materials and Technologies*, Universita Politecnica delle Marche, Ancona, Italy, June 28–30, 2010.

30. Shima H., Nakato, T.H., Okamoto, M., Asano, T., et al., New technology for recovering high quality aggregate from demolished concrete. *Proceedings of Fifth International Symposium on East Asia Recycling Technology*, 1999.

31. Akbarnezhad, A., Ong, K.C.G, et al., Microwave-assisted beneficiation of recycled concrete aggregates. *Construction and Building Materials*, 2011, **25**:3469–3479.

32. Homand-Etienne, F. and R. Houper, Thermally induced microcracking in granites: characterization and analysis. *International Journal of Rock Mechanics and Mining Sciences and Geomechanics Abstracts*, 1989, **26**(2):125–134.

33. Yoda, K., Harada, M., and Sakuramoto, F., *Field Application and Advantage of Concrete Recycled In-Situ Recycling Systems*. London: Thomas Telford Services, 2003.

34. Mindess, S., Young, J.F., and Darwin, D., *Concrete*, 2nd ed. Upper Saddle River, NJ: Pearson Education.

35. Tam, V.W.Y., Tam, C.M., and Le, K.N., Removal of cement mortar remains from recycled aggregate using pre-soaking approaches. *Resources, Conservation and Recycling*, 2007, **50**(1):82–101.

36. Sheppard, W.L., Jr., *Corrosion and Chemical Resistant Masonry Materials Handbook*. Saddle River, NJ: Noyes, 1986.

37. Kessler, D.W., Insley, H., and Sligh, W.H., *Physical Mineralogical and Durability Studies on Building and Monumental Granites of the United States*. Research Paper 1320. Washington, DC: National Bureau of Standards, 1940.

38. Lippiatt, N. and Bourgeois, F., Investigation of microwave-assisted concrete recycling using single-particle testing. *Minerals Engineering*, 2012, **31**:71–81.

39. EnviroCentre, Controlled Demolition, National Green Specification. *Demolition: Implementing Best Practice*. Banbury, UK: Waste and Resources Action Programme, 2005.

Chapter 5

Process control in microwave heating of concrete

5.1 INTRODUCTION

A number of potential applications of microwave heating in the production, demolition, and recycling of concrete were discussed in the previous chapters. In Chapter 2, we observed that uniform microwave heating at a lower industrial, scientific, and medical (ISM) microwave frequency range may be used to accelerate the curing of concrete. However, we also observed that the uniformity of microwave heating is a key factor affecting curing quality. Differential heating of concrete components as a result of the selection of an inappropriate microwave frequency or inappropriate design of the microwave applicator may result in the development of undesirable thermal stresses within the cured concrete components. Such differential stresses can negatively affect the mechanical properties and durability of the concrete components being cured by contributing to the growth of microcracks present in concrete.

To achieve a uniform microwave heating pattern, besides choosing an appropriate microwave frequency, the microwave-curing chamber should also be properly designed. However, there are as yet no guidelines in the selection of reliable design techniques to achieve uniformity of microwave power applied across a typical microwave-curing chamber, and the few designs that have been adopted are usually based on the engineer's or practitioner's experience. Hence, a process control technique is required to monitor and ensure uniformity of microwave heating throughout the curing process.

In addition to the uniformity of heating requirement, an important requirement to ensure optimal microwave curing of concrete components is to maintain a level of heating that is optimal throughout the microwave-curing process. It has been reported that overheating during curing may negatively affect the long-term properties of the microwave-cured concrete [1]. However, maintaining the concrete temperature at the desired range during the microwave-curing process is difficult and requires continuous monitoring and feedback control. This is because the temperature of the concrete being microwave heated at a particular microwave frequency and power is controlled by the size of the concrete component and the dielectric properties of the concrete present. The size and dielectric properties of concrete components fabricated in a concrete prefabrication plant may vary considerably depending on the type and number of precast concrete components produced at any particular time. In addition, the dielectric properties of the concrete may vary considerably during the microwave-curing process with the changes in the concrete's water content. Unless the components are sealed completely, the water content of the concrete being microwave cured is likely to reduce with time as the available free water present in the concrete evaporates as a result of the microwave heating.

With this in mind, continuous fine-tuning and feedback adjustments of microwave power may be required to avoid localised overheating and underheating throughout the concrete mass or volume during the microwave-curing process. Therefore, a monitoring and feedback control mechanism to ensure the essential optimal level and pattern of heating throughout the curing process is necessary to guarantee the quality of the microwave-cured precast concrete products. Undoubtedly, the simplest and most logical way to control a heating process is via temperature monitoring. Leung and Pheeraphan showed that an optimal microwave-curing regime that provides high strength at both the early and later stages can be achieved with the help of a temperature control system [2]. Figure 5.1 illustrates how temperature monitoring can be used as a feedback control mechanism for the accelerated microwave curing of concrete.

In addition, in Chapter 3, we observed that in microwave-assisted selective demolition of concrete, the thermal stresses and pore pressure developed in the concrete component may vary as a function of temperature and its gradient. This suggests that monitoring the temperature of concrete may serve as a simple means to monitor the thermal stresses and pore pressure developed within the concrete and that temperature feedback control may be used effectively in microwave-assisted selective demolition of concrete. The surface temperature of the concrete component being heated using a microwave-assisted selective demolition tool may be used to predict the heating duration needed to achieve the desired removal depth. Accordingly, based on real-time feedback, the microwave power can be adjusted continuously to minimise the differences between the actual and planned removal depths and rates. Figure 5.2 illustrates how temperature monitoring can be used as a feedback control mechanism for microwave-assisted selective concrete demolition systems.

In Chapter 4, a microwave separation method for reducing the mortar content of recycled concrete aggregates (RCAs) through partial or complete removal of the mortar adhering to the RCA particles was introduced. We observed that the separation of mortar is achieved through development of relatively high differential thermal stresses in the adhering mortar and its

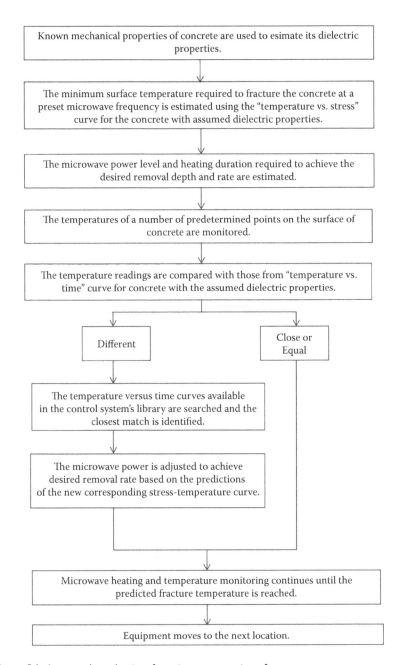

Figure 5.1 A control mechanism for microwave curing of concrete.

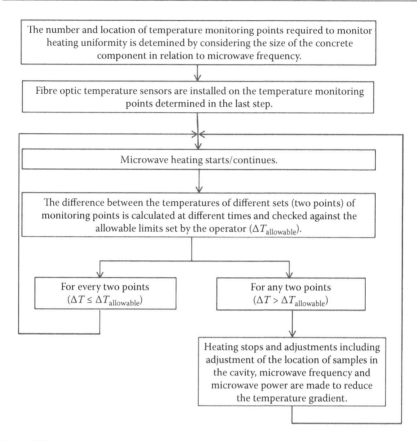

The number and location of temperature monitoring points required to monitor heating uniformity is detemined by considering the size of the concrete component in relation to microwave frequency.

Fibre optic temperature sensors are installed on the temperature monitoring points determined in the last step.

Microwave heating starts/continues.

The difference between the temperatures of different sets (two points) of monitoring points is calculated at different times and checked against the allowable limits set by the operator ($\Delta T_{allowable}$).

For every two points
($\Delta T \leq \Delta T_{allowable}$)

For any two points
($\Delta T > \Delta T_{allowable}$)

Heating stops and adjustments including adjustment of the location of samples in the cavity, microwave frequency and microwave power are made to reduce the temperature gradient.

Figure 5.2 A feedback control mechanism for microwave-assisted selective demolition of concrete.

interface with the natural aggregates in the RCA. The temperature required to fracture or detach the adhering mortar in this method is in the range of 150°C–200°C, depending on the strength of mortar. Temperature monitoring during the microwave separation process is crucial to avoid underheating or overheating of the RCA, thus ensuring the optimal quality of RCA produced. Although underheating can affect the efficiency of the mortar removal process, overheating may have considerable negative effects on the properties of the natural aggregates present in the RCA. Additional unnecessary costs and undesirable environmental impact may be introduced if the microwave separation process is inefficient. Adjusting the microwave power and heating duration based on temperature feedback can be used as a possible feedback control mechanism to avoid RCA overheating or underheating during the microwave separation process.

Similar to the microwave-assisted methods described so far, temperature monitoring is the most common process control methodology adopted in almost all microwave-processing applications. Inaccuracies in temperature measurements during monitoring can lead to erroneous indications of the process temperature and provide misleading representations of the process efficiency. However, despite its importance, temperature is one of the most difficult parameters to measure in a microwave environment. In the following section, the suitability of some of the most common means of temperature monitoring, including thermocouples, infrared thermal cameras, and fibre-optic sensors, for applications in the microwave processing of concrete is discussed.

5.2 TEMPERATURE MEASUREMENT IN MICROWAVE HEATING OF CONCRETE

In the microwave heating of concrete, temperature measurements have to be made directly on or within the sample, not merely near the vicinity of the microwave applicator. This is because microwaves heat the sample volumetrically, not just its surroundings. Hence, any temperature probe used must maintain good contact with the sample for accurate temperature measurements. In the following, some of the most common temperature-monitoring probes are introduced and their suitability for temperature measurements in microwave heating applications is discussed.

5.2.1 Thermocouples

Thermocouples are the most commonly used temperature measurement sensors. A thermocouple consists of two dissimilar metals joined together at one end (Figure 5.3). When the junction of the two metals is heated or cooled, a voltage is produced that can be correlated back to the temperature at the location where the thermocouple is located. The thermocouple alloys are commonly available as wires. Thermocouples are available in

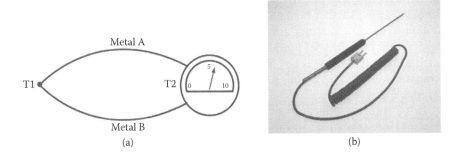

Figure 5.3 (a) Basic thermocouple circuit and (b) a commercially available thermocouple.

different combinations of metals and calibrations. The four most common calibrations are graded as types J, K, T, and E. There are also high-temperature calibrations R, S, C, and GB. Each type (or calibration) has a different temperature range of applicability, although the maximum temperature that may be monitored also varies with the diameter of the wire used in the thermocouple.

Despite the popularity of thermocouples in many civil engineering applications, they are not suitable as probes for temperature monitoring in most microwave-assisted processes. Because of their metallic nature, thermocouples are not immune to electromagnetic interference and may be inaccurate when utilised in a microwave field. The presence of an electrically large ($>\lambda/10$, where λ is the wavelength of the electromagnetic wave) metallic temperature probe in a microwave environment may cause localised distortions in the electromagnetic field distribution and induce electrical currents that can affect the electronics used for temperature measurements. The presence of an electromagnetic field may also give rise to errors caused by self-heating, heat conduction, or excessive localised heating, particularly at the tip of the temperature probe.

5.2.2 Infrared thermal cameras (radiation thermometry)

Infrared thermography converts heat at any temperature into a thermal image using specialised scanning equipment [3]. Radiation is the mode of heat transfer that easily allows the recording of a visual image of the thermal conditions prevailing. Thermography senses the emission of the thermal radiation from the material surface being observed. Radiation can result from either active or passive sources [4]. A surface of temperature T emits infrared radiation of intensity $I = A\ T^4$, where A is a constant. Radiometric measurements of I serve as an accurate indication of T. In thermal cameras, the infrared radiation emitted from the measuring object is detected and converted to an electric signal by the two-dimensional uncooled focal plane array detector, and the amplitude analog temperature signal is converted into a digital signal. The digital signal is displayed as a thermal image in colour or black and white. Infrared imaging is widely used for nondestructive testing of industrial products, in aircraft surveillance, and for medical diagnostics. It has also been used for quick characterisation of microwave components [5]. Numerous infrared thermal cameras are commercially available (Figure 5.4).

In addition to applications in temperature monitoring to avoid overheating and underheating, infrared thermography can serve as an excellent method for the detection of cracks developing on the surface of the concrete components processed with microwaves. This can be useful especially in accelerated microwave curing of concrete, for which development of thermal

Figure 5.4 An infrared thermal camera.

cracks can negatively affect the mechanical properties and long-term durability of the concrete. Infrared thermography crack detection utilises the principle that a defect restricts heat transfer into or out of a material, as an air gap is introduced that acts as an insulator at the defect zone. This may be used to detect the presence of surface cracks. In addition, differences in thermal conductivity, specific heat capacity, or density at different locations on the surface result in measureable temperature differences that can be used to assess structural nonhomogeneity at a particular location [6]. That is, defect-free concrete surfaces will appear uniform in colour and texture when viewed using an infrared camera. Cracks or delaminations within the concrete will heat at a faster rate and will be observed on the thermal image as hot spots (Figure 5.5). This method is therefore well suited for the rapid assessment of large concrete surfaces [3].

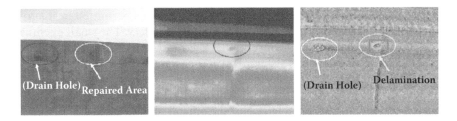

Figure 5.5 Infrared thermography of a concrete bridge deck. (From Mitani, M., Matsumoto, K., and Necati Catbas, F., Innovative bridge assessment methods using image processing and infrared thermography technology, paper presented at the 37th Conference on Our World in Concrete and Structures, August 29–31, 2012, Singapore. With permission).

When applying infrared thermography to detect surface defects, the effects of convection and conduction should be taken into account. Convection is a significant mode of heat transfer, especially if measurements are made in the field. This is because fluid flow along the surface can transfer heat. Wind velocity is therefore a critical parameter that affects the precision of this method. Other critical parameters include solar radiation, surface emissivity, and ambient temperature. If not carefully controlled, these factors may complicate data interpretation [4]. Conduction may result in surface defects that are small in size compared to their depth and be undetectable with thermography. This is because energy is gradually diffused into the material with little thermal contrast observed at the surface [4]. For this reason, thermography is more suited to detecting the presence of larger surface defects. Thermography cannot provide data on the depth of defects; however, if combined with radar (microwave) techniques, this limitation can be overcome.

5.2.3 Fibre optics

When temperature measurements are performed in the presence of an electromagnetic field, conventional temperature sensors such as thermocouples may not be accurate. This is mainly because of the metallic nature of these sensors. The induced currents and voltages in the metallic conductors and the local heating of the sensor as a result of electromagnetic induction cause electromagnetic interference, reducing the precision of the metal-based sensors [7]. To improve the accuracy of temperature measurement in the microwave processing of materials, temperature sensors based on optical fibres have been studied.

Optical fibre sensors are well known to be virtually unaffected by electromagnetic fields [7–10]. Because of their glass-based nature, optical fibres do not interact with the electromagnetic field and hence maintain their accuracy in the presence of microwaves. The considerably smaller size of optical fibres compared to conventional temperature sensors facilitates the wiring needed for instrumentation inside the microwave heating chamber. This is especially important because the size of the openings permitted in microwave heating chambers is usually rather small for safety. The small opening makes it difficult to deploy conventional thermocouples for measurements.

The use of optical fibres as temperature sensors has been investigated by several researchers in different fields [7,11–24]. The more commonly used types of optical fibre temperature sensors may be categorised into three groups. In the first group, a length of the optical fibre's cladding is replaced with materials that have a temperature-dependent refractive index [7]. The variations of the refractive index in such sensors leads to variations in the optical power transmitted along the fibre, which is then correlated to indicate temperature changes. The second group of optical fibre temperature sensors makes use of the fluorescence lifetime approach. Such sensors have been reported to work satisfactorily in conditions ranging from below room temperature to above 300°C [25]. Finally, the last group of optical fibre sensors, which are referred to as fibre Bragg grating (FBG) sensors, are made by forming a distributed Bragg reflector within a short segment of an optical fibre that reflects particular wavelengths and transmits all others. FBG sensors are considered one of the most suitable temperature sensors for temperature measurement in the presence of strong electromagnetic fields. Because of their suitability for the applications discussed previously in this book, the working principles of this group of optical fibre sensors are described in detail in the following sections.

5.2.3.1 Fibre Bragg grating sensors: Structure and working principles

The FBGs have been used as temperature sensors in many civil engineering applications, mostly related to the monitoring of temperature of the concrete subjected to elevated temperatures in the absence of strong electromagnetic fields [19,21,26,27]. The satisfactory employment of FBGs in noncivil engineering applications involving the use of strong electromagnetic fields has also been reported by a number of studies [7,9,17,28]. The results of a recent study showed that certain types of FBG sensors can be used as temperature sensors in the microwave-assisted applications discussed in Chapters 2 to 4. The findings of this study are reviewed and discussed in detail in the subsequent sections. However, prior to illustrating the suitability and precision of FBGs as temperature sensors for microwave-assisted applications, a brief introduction to the working principles of FBGs is first provided to help readers appreciate what makes FBGs good temperature sensors in the presence of electromagnetic fields.

A Bragg grating is a periodic structure fabricated by exposing parts of a photosensitised fibre core to ultraviolet light (Figure 5.6) [29]. Figure 5.7 shows a typical commercially available FBG sensor. When light propagating along a single-mode fibre encounters the periodic variation in the refractive index of the in-fibre grating, a small amount of light is reflected at each point of the refractive index transition. If each of the reflections is in phase, they will add coherently and produce a large net reflection from the grating. This phase matching occurs typically at only one specific wavelength, and this wavelength is called the Bragg wavelength. Any thermal expansion, strain, or pressure change in the structure monitored by the FBG leads to a change in the grating spacing and thus a change in the FBGs' refractive index, which results in a shift in the Bragg wavelength [30]. The Bragg wavelength, λ_B, is related to the grating period and the effective refractive index of the fibre n_{eff} by

$$\lambda_B = 2 \Lambda n_{eff} \tag{5.1}$$

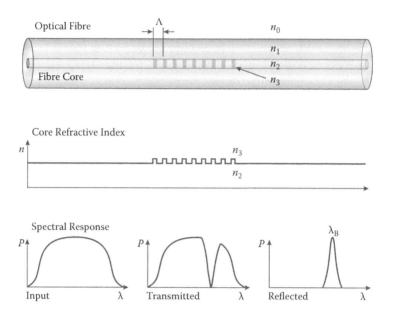

Figure 5.6 A fibre Bragg grating structure with refractive index profile and spectral response.

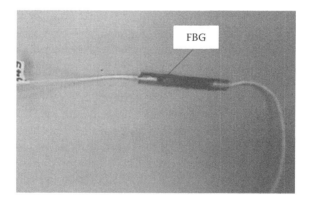

Figure 5.7 A commercially available FBG sensor.

Both the effective refractive index and the grating period vary with changes in strain ε, temperature T, and pressure P imposed on the fibre. An applied longitudinal strain or pressure (radial strain) will shift the Bragg wavelength through expansion or contraction of the grating periodicity and through photoelastic effects. Temperature affects the Bragg wavelength through thermal expansion and contraction of the grating periodicity and through the thermal dependence of the refractive index. These effects are well understood and, when adequately modelled, provide a means for measuring strain, pressure, and temperature. The shift of the Bragg wavelength with temperature and strain is given by the following expression:

$$\Delta\lambda_B = 2n_{eff}\Lambda\left[(1 - P_e)\Delta\varepsilon + (\alpha + \xi)\Delta T\right] \tag{5.2}$$

where P_e is the effective elastic-optic coefficient, $\Delta\varepsilon$ is the applied strain change, α is the thermal expansion coefficient, ξ is the thermo-optic coefficient, and ΔT is the temperature change. The strain response depends on the physical elongation of the sensor and the change in the fibre index because of photoelastic effects. The temperature response depends on the fibre material's thermal expansion and the temperature dependence of the refractive index because of thermo-optic effects. Such a thermally induced Bragg wavelength shift is given by [15]

$$\Delta\lambda_B = 2\Delta n\Lambda + 2n\varepsilon_2\Lambda = 2n\Lambda\left\{\frac{\partial n}{\partial T}\frac{1}{n}\Delta T + \varepsilon_z - \frac{n^2}{2}(P_{11} + P_{12})\varepsilon_r + P_{12}\varepsilon_z\right\} \tag{5.3}$$

where ε_z and ε_r represent the axial strain and the radial strain in the core, respectively, and P_{11} and P_{12} are the Pockels coefficients of the core [14]. According to Equation 5.1, if the lateral and longitudinal strains of the fibre are set to zero, temperature would be directly related to the change in the wavelength:

$$\Delta\lambda_B = 2\Delta n\Lambda + 2n\varepsilon_z\Lambda = 2n\Lambda\frac{\partial n}{\partial T}\frac{1}{n}\Delta T = K_T\Delta T \tag{5.4}$$

The lateral and longitudinal displacement freedom of FBGs can be provided through the design of the sensor's coating or by avoiding constraints when installing the FBG sensors on the structures to be instrumented. Equation 5.4 clearly illustrates the concept behind an FBG temperature sensor. The coefficient K_T may be obtained through calibration.

Many researchers have tried to increase the sensitivity of the FBG temperature sensors. Generally, two techniques have been used for this purpose. In the first technique, the FBG sensors are coated using different coatings with higher expansion coefficients than the bare FBG itself [14,31]. By using such coatings, a given temperature change will translate to a higher thermal expansion of the FBG and hence will increase the sensitivity of the FBG sensors.

In the second technique, two metallic pieces are connected to both ends of the FBG sensors and the difference in the expansion of the two metals possessing different thermal expansion coefficients is translated into a temperature difference [18].

5.2.3.2 Fibre Bragg grating sensors in microwave fields

Unlike conventional heating, microwaves heat the sample itself volumetrically and not the surroundings. Hence, if the fibre's coating is made of a microwave-absorbent material, it will itself be heated; as a result, the optical fibre would measure the temperature of its own coating rather than the temperature of the material that is meant to be monitored. With this in mind, the optical fibre sensors used for temperature measurement in an electromagnetic field should not have coatings unless the coating is transparent to electromagnetic waves. Moreover, when selecting optical sensors for temperature measurements in the microwave-assisted applications discussed in this book, optical fibres with a metallic coating or those based on the differential expansion of two metallic pieces at the FBG ends should be avoided. This is again because of the possible interference caused by currents and voltages induced in the metallic conductors and the local heating of the sensor itself because of electromagnetic induction [7].

5.3 TEMPERATURE MONITORING OF MICROWAVE-ASSISTED CONCRETE PROCESSES

The three most common types of temperature sensors used in process feedback control in different industries were introduced in the previous sections. In this section, the suitability of these sensors for temperature monitoring in the microwave-assisted processes used in concrete production, demolition, and recycling is discussed. A number of important considerations to be taken into account at the various stages of selection, calibration, and installation of FBG sensors as well as the data collection and analysis stage are discussed. The results of a series of experimental and numerical studies performed using various temperature sensors discussed previously are used to illustrate the precision of FBG sensors in a microwave field.

The microwave oven used in these experiments consisted of two magnetrons, each with a maximum power of 950 W. The interior of the microwave oven is shown in Figure 5.8. As shown, the microwaves generated are transferred to the microwave cavity through two WR340 rectangular waveguides. As can be seen in Figure 5.8, the microwave oven originally used included a mode stirrer to help increase the uniformity of the microwave power inside the cavity. Because of the difficulties in the numerical simulation of the behaviour of the stirrer, the stirrer was removed during the experiments so that the results of the experimental study could be compared with the results of the numerical simulation.

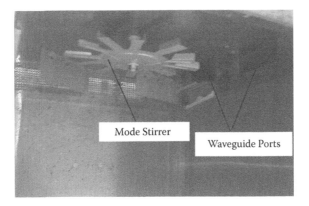

Mode Stirrer

Waveguide Ports

Figure 5.8 Interior of the microwave oven.

5.3.1 Selection and calibration of FBG sensors

To compare the accuracy of different temperature sensors (thermocouples, fibre optics, and thermal cameras) in a microwave field, a fair comparison basis must be ensured by selecting the most appropriate temperature sensors for each type. With this in mind, selection of a suitable FBG sensor is crucial. This is because of the wide range of FBG sensors available commercially. It should be noted that many of the commercially available optical fibre temperature sensors comprise a metallic shielding or use metallic fibre splicers that are undesirable for use in microwave-assisted applications. Moreover, as discussed, the working mechanism of some FBGs is based on correlating the relative difference in the strain of two metals attached to each other, exposed to heating [14,18,31]. Because of the electromagnetic interference caused by the metallic components, such sensors should be avoided when selecting FBGs for temperature measurements in the presence of a microwave field.

A commercial optical fibre sensor with embedded metallic fibre splicers heated in a commercial microwave oven is shown in Figure 5.9. The plastic coating next to the metallic splicers began to ignite after a few seconds of microwave heating as it absorbed microwave energy. As the metallic splicer was exposed to microwave radiation directly, resulting in the loss of the plastic coating, the induction phenomenon led to sparks generated inside the microwave chamber. This finally led to the splitting of the fibre next to the metallic splicer. This clearly illustrates the importance of selecting an appropriate FBG sensor for monitoring the temperature of concrete in microwave-assisted concrete-processing applications.

In addition to the problems associated with the presence of metallic components within the FBG sensors, other issues related to the microwave power absorption of the fibre covers should be taken into consideration

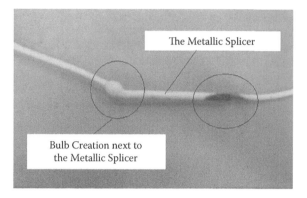

Figure 5.9 Combustion of the plastic coating next to the metallic splicer after microwave heating.

when selecting FBG sensors for temperature measurements in the presence of a microwave field. The coated optical fibres usually used for strain and temperature measurements in civil engineering applications are generally not suitable for temperature monitoring in a microwave field. This is because microwaves may heat the coating to different extents compared to the concrete specimen, depending on the difference between the dielectric properties of the coating material and concrete.

In summary, to select an appropriate FBG type for use in microwave-assisted concrete processing applications, the following two important points should be taken into account:

1. The FBG sensor should not contain any metallic components, whether in the form of fibre splicers, within the cover, or as part of the sensor's structure.
2. The covering on the fibre should be derived from microwave-transparent materials. Bare fibres are suitable but should be handled with care because of their fragile nature.

After selection of the most suitable FBG sensors and prior to installation of sensors on the concrete component, FBG sensors should be calibrated. The calibration process allows for the determination of the thermal-optic coefficient K_T of the sensor. The calibration may be performed by placing FBGs together with two or more thermocouples in a conventional oven. The change in the wavelength of the fibres and the temperatures monitored by the thermocouples should then be recorded using an FBG interrogator system and a data logger, respectively, as the oven's temperature is increased from room temperature, say 20°C, to the estimated maximum temperature measurable with the FBG. The use of multiple thermocouples

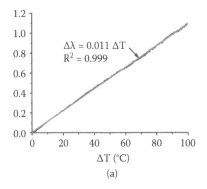
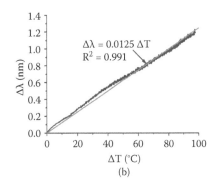

$\Delta\lambda = 0.011 \Delta T$
$R^2 = 0.999$

$\Delta\lambda = 0.0125 \Delta T$
$R^2 = 0.991$

Figure 5.10 Calibration curves for the two bare FBG sensors (FBG1 (a) and FBG2 (b)) used in the examples presented in this chapter.

is to reduce potential measurement errors. The calibration curves of two bare FBGs used in the illustrative examples presented in this chapter are shown in Figure 5.10.

5.3.2 Installation of sensors and temperature monitoring during microwave heating

To compare the accuracy of different temperature sensors in the presence of a microwave field, a case study is described. Three T-type thermocouples capable of measuring temperatures up to 250°C were embedded at different depths inside the concrete specimens during casting to monitor the internal temperature change. Moreover, two FBG sensors were mounted on the surface of the specimens at the time of testing to measure the surface temperature. The placement of thermocouples inside the concrete specimens was to minimise exposure of the thermocouple's metallic head to the electromagnetic field while the concrete specimens were being microwave heated. This is because exposure of the metallic head of a thermocouple leads to the occurrence of induction phenomenon between the thermocouple's head and the microwave cavity, which is likely to lead to erroneous readings. Moreover, the installation of FBGs on the surface rather than within the concrete specimens was because of the need to verify the accuracy of the FBG sensors using another accurate and reliable means of instrumentation, which in the case of these experiments was via a noncontact infrared thermal camera. The infrared thermography cameras can only capture the surface temperature. Hence, for comparison purposes, FBGs were used to monitor the surface temperatures for comparison with those monitored by the infrared camera. The location of the various sensors in the specimens tested is shown schematically in Figure 5.11.

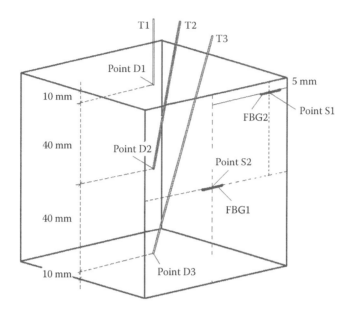

Figure 5.11 Instrumentation of concrete specimens: positioning of thermocouples and FBG sensors.

The concrete specimens used were all in a saturated surface dry condition and 70 days old at the time of testing. Concrete specimens were heated at two different power levels of 950 and 1900 W for 2 minutes. The cavity's door was immediately opened after heating, and the temperature of the front surface of the concrete specimen (also monitored by FBGs) was captured using the infrared thermal camera. The surface temperature profile of the specimens heated at 950 and 1900 W are shown in Figures 5.12 and 5.13, respectively.

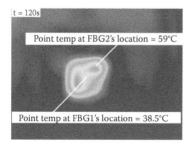
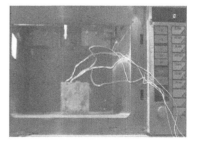

Figure 5.12 The surface temperature profile of the concrete specimen heated at 950 W for 2 minutes.

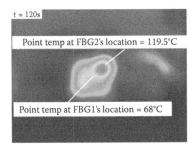

Figure 5.13 The surface temperature profile of the concrete specimen heated at 1900 W for 2 minutes.

5.3.3 Estimation of sensors' temperature readings using numerical simulation

Numerical simulation can be used to provide another basis for verification of the accuracy of the temperature measurements [32,33]. To simulate the microwave heating process and estimate the temperature distribution in the concrete specimens heated, the exact configuration of the microwave oven and specimens should be modelled. This is because, as is discussed in detail in Chapter 6, the configuration of the microwave cavity and concrete specimen and the placement of the latter in the microwave cavity can considerably affect the resulting heating pattern. A sketch of the microwave oven used in the numerical study conducted on our example case is shown in Figure 5.14. The heating process was simulated using the electromagnetic module of the COMSOL Multiphysics commercial finite-element package. The analytical modelling of the microwave heating process involves solving Maxwell's equations, which govern the propagation of microwave radiation through concrete and the microwave's waveguide or cavity, and the heat transfer equation, which governs heat absorption and the resulting temperature rise within the concrete block.

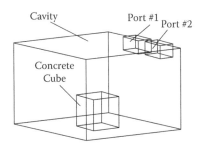

Figure 5.14 Sketch of the microwave oven for numerical modelling.

5.3.4 Precision of thermocouples

The temperature measurements using thermocouples for six different concrete specimens are shown in Table 5.1. As shown, the thermocouples used for temperature monitoring in the microwave field were either dysfunctional or inaccurate. Table 5.1 shows that in three out of the six concrete specimens tested, at least one of the embedded thermocouples was totally damaged during the test or that the temperature monitored was obviously erroneous or out of range. Therefore, conventional thermocouples are not reliable as temperature sensors for temperature monitoring in microwave-assisted processing of concrete.

Figures 5.15 and 5.16 present the temperature measurement results for two of the three cases in which all thermocouples were functional during microwave heating. In these figures, the thermocouple readings at two different microwave power levels of 950 and 1900 W are compared with the respective temperatures predicted through numerical simulation. The results are compared over a period of 4 minutes, including 2 minutes of microwave heating and 2 minutes of cooling.

As shown in Figures 5.15b and 5.16b, according to the numerical simulation results, microwave heating is expected to result in a linear increase in the concrete temperature until a maximum is reached at the end of the heating period, while the rate of temperature rise is proportional to the microwave power. During the cooling period, as a result of heat transfer from the hotter locations to the cooler locations, the temperature at each location tends to converge to a stable average value. As shown in Figures 5.15 and 5.16, considerable differences were observed between the temperature readings of thermocouples during the heating period and the respective temperatures predicted through numerical simulation. Such a considerable difference clearly shows that the thermocouples' readings during the microwave heating period (from 0 to 120 s) were erroneous. The

Table 5.1 Thermocouple readings at $t = 120$s

Specimen	Microwave power (W)	Thermocouple readings (°C)			Analytically predicted temperatures (°C)		
		DI	D2	D3	DI	D2	D3
CI	900	77.0	DF[a]	80.0	77.5	67.9	70.5
C2	1800	140.1	90.4	116.0	134.5	90.9	117.6
C3	900	72.7	78.8	68.0	77.5	67.9	70.5
C4	1800	140.0	96.8	122.6	134.5	90.9	117.6
C5	900	DF	82.7	DF	77.5	67.9	70.5
C6	1800	174.3	DF	127.3	134.5	90.8	117.6

[a] DF: dysfunctional; that is, the thermocouple was damaged as a result of the electromagnetic interference.

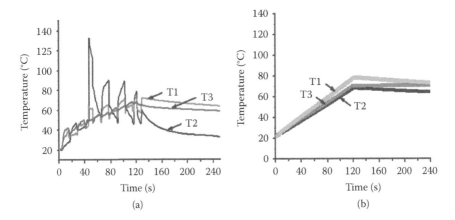

Figure 5.15 Temperatures at locations monitored using embedded thermocouples for a saturated concrete specimen heated at 950 W for 2 minutes: (a) actual readings of embedded thermocouples; (b) temperatures predicted using numerical simulation.

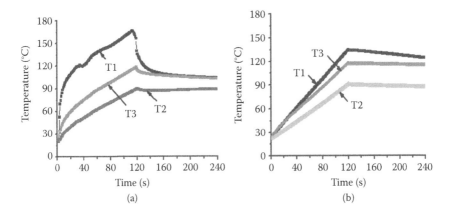

Figure 5.16 Temperatures at locations monitored using embedded thermocouples for a saturated concrete specimen heated at 1900 W for 2 minutes: (a) actual readings of embedded thermocouples; (b) temperatures predicted using numerical simulation.

thermocouple measurements were nonlinear and fluctuated in some cases (e.g., the T2 thermocouple in Figure 5.15a).

The inaccurate thermocouples' measurements observed in this illustrative example highlight the disruptive effects of the strong electromagnetic field present in the microwave oven on the accuracy of the thermocouples. This confirms that conventional thermocouples are not suitable as temperature sensors for process monitoring and feedback control in microwave-assisted processing of concrete.

5.3.5 Precision of FBGs

Figures 5.17 and 5.18 compare the temperature measurements recorded using the FBG sensors to the respective temperatures predicted using numerical simulation of the microwave heating process. In addition, the temperature measurements at the locations where the FBGs were mounted,

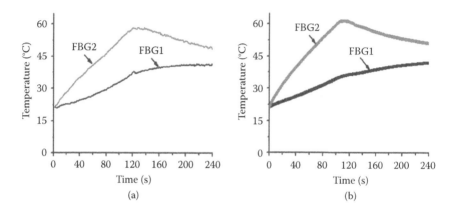

Figure 5.17 Temperatures at locations where FBGs are mounted for a saturated concrete specimen heated at 950 W for 2 minutes: (a) actual readings of FBGs; (b) temperatures predicted using numerical simulation.

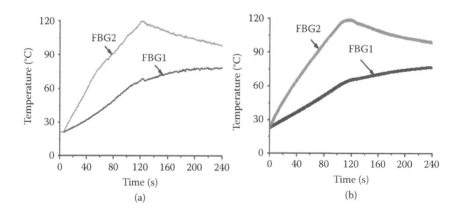

Figure 5.18 Temperatures at locations where FBGs are mounted for a saturated surface dry concrete specimen heated at 1900 W for 2 minutes: (a) actual readings of FBGs; (b) temperatures predicted using numerical simulation.

Table 5.2 Temperature after two minutes of microwave heating at the locations where FBG sensors are mounted, as measured by the infrared thermal camera and FBGs sensors and predicted using numerical simulation

Specimen	Microwave Heating Power (W)	Thermotracer Readings at t = 120 s (°C)		FBG Readings (°C)		Analytically Predicted Temperatures (°C)	
		S I	S2	S I	S2	S I	S2
C I	950	39.0	57.5	39.0	58.3	36.2	60.9
C2	1800	67.0	119.5	67.2	119.2	65.3	119.0
C3	950	38.5	59.0	37.6	58.2	36.2	60.9
C4	1800	66.5	120.0	65.7	120.4	65.3	119.0
C5	950	36.0	118.0	35.6	117.8	36.2	60.9
C6	1800	70.0	124.0	70.0	123.7	65.3	119.0

those obtained using the infrared thermal camera, and those predicted through numerical simulation are compared in Table 5.2. As shown in Figures 5.17 and 5.18, FBGs and numerical simulation indicated almost similar temperature development trends and peak temperatures at the end of the heating period. Similarly, the results of our case study show good consistency and close agreement between the temperature measurements during the cooling period obtained using FBGs and those predicted using numerical simulation.

For both FBG sensors (FBG1 and FBG2), the experimental measurements and numerical predictions showed an almost linear increase in the surface temperature of concrete for the entire duration of the microwave heating period. Again, the temperature measurement curves associated with FBG1 and FBG2 converged during the cooling period as heat was transferred from hotter locations (measured by FBG2) to cooler locations (measured by the FBG1).

It should also be noted that the small differences observed between the FBG readings and the results of numerical simulations are likely to be partially attributable to the differences between the actual magnitude of the electromagnetic properties of concrete and those assumed in the numerical simulation. As discussed in Chapter 1, each concrete specimen has a unique set of electromagnetic properties determining the amount of energy absorbed, reflected, and transferred when it is exposed to microwaves. Accurate measurement of the electromagnetic properties is usually performed using a network analyser comprising a dielectric measurement kit. However, in the illustrative example presented in this chapter, rather than using the actual electromagnetic properties of each concrete specimen tested, the average of the dielectric properties of typical saturated surface dry concrete were used. This probably accounts for the minor discrepancies observed.

The accuracy of the FBG measurements can also be verified using the surface temperatures recorded by the infrared thermal camera. The temperatures at the same locations monitored by the FBG sensors were compared with those measured using the infrared camera and are shown in Figures 5.12 and 5.13. A comparison of the maximum temperatures measured by the FBG sensors and the temperatures at the same locations monitored using the infrared thermal camera again confirm that the accuracy of the bare FBG sensors is satisfactory for use in processes associated with the microwave heating of concrete.

5.4 SUMMARY

Temperature is one of the most important process-monitoring and feedback control measures for microwave-assisted concrete processing. However, accurate measurement of temperatures in the presence of a strong electromagnetic field is difficult. This is mainly because most conventional metal-based temperature sensors are either dysfunctional or lack precision when used in an electromagnetic field. In this chapter, the suitability of some of the more common temperature sensors for use in temperature monitoring and feedback control in microwave-assisted processing of concrete was discussed. It was noted that conventional thermocouples are not suitable as sensors in microwave-processing applications as the readings of the thermocouples in the presence of an electromagnetic field are not accurate and mostly overestimate the actual temperature of the concrete. On the contrary, we observed that the temperature of concrete in a microwave field can be accurately monitored using bare FBG sensors. Therefore, bare FBG sensors may be used as reliable process feedback control tools in microwave-processing applications. However, it should be noted that commercially packaged, off-the-shelf FBG sensors currently used for strain and temperature measurements of concrete in normal field environments cannot be used in the presence of strong microwave fields. This is mainly because of the self-heating of the outer polymer sheath present in such FBG sensors. Also, there is uncontrolled electromagnetic interference caused by the metallic connectors present in such optical fibres.

REFERENCES

1. Leung, C.K.Y. and Pheeraphan, T., Very high early strength of microwave cured concrete. *Cement and Concrete Research*, 1995, 25(1):136–146.
2. Leung, C.K.Y. and Pheeraphan, T., Determination of optimal process for microwave curing of concrete. *Cement and Concrete Research*, 1997, 27(3):463–472.

3. McCann, D.M. and Forde, M.C., Review of NDT methods in the assessment of concrete and masonry structures. *NDT & E International*, 2001, **34**(2):71–84.

4. Buyukozturk, O., Imaging of concrete structures. *NDT & E International*, 1998, **31**(4):233–243.

5. Kiwamoto, Y., Abe, H., et al., Thermographic temperature determination of gray materials with an infrared camera in different environments. *Review of Scientific Instruments*, 1997, **68**(6):2422.

6. Maierhofer, C., Brink, A., et al., Detection of shallow voids in concrete structures with impulse thermography and radar. *NDT & E International*, 2003, **36**(4):257–263.

7. Pennisi, C.P.A., Leija, L., et al., Fiber optic temperature sensor for use in experimental microwave hyperthermia. *Proceedings of IEEE Sensors*, 2002, **2**:1028–1031.

8. Schmaup, B., Mirz, M., and Ernst, J., A fiber-optic sensor for microwave field measurements. *Review of Scientific Instruments*, 1995, **66**(8):4031–4033.

9. Degamber, B. and Fernando, G.F., Fiber optic sensors for noncontact process monitoring in a microwave environment. *Journal of Applied Polymer Science*, 2003, **89**:3868–3873.

10. McSheny, M., Fitzpatrick, C., and Lewis, E., Investigation and development of a fibre optic temperature sensor for monitoring liquid temperature in a high power microwave environment. *Proceedings of SPIE 5502, Second European Workshop on Optical Fibre Sensors*, June 9, 2004, 80–83.

11. Kersey, A.D. and Berkoff, A., Fiber-optic Bragg-grating differential-temperature sensor. *IEEE Photonics Technology Letters*, 1992, **4**(10):1183–1185.

12. Jaehoon Jung, Hui Nam, Byoungho Lee, Jae Oh Byun, and Nam Seong Kim, Fiber Bragg grating temperature sensor with controllable sensitivity, *Applied Optics*, 1999, 38(13):2752-2754. http://dx.doi.org/10.1364/AO.38.002752

13. Silva, J.C.C., Martelli, C., et al., Temperature effects in concrete structures measured with fibre Bragg grating. *Proceedings of SPIE 5502, Second European Workshop on Optical Fibre Sensors*, June 9, 2004, 68–71.

14. Seo, C. and Kim, T., Temperature sensing with different coated metals on fiber Bragg grating sensors. *Microwave and Optical Technology Letters*, 1999, **21**:162–165.

15. Lagakos, N., Bucaro, J.A., and Jarzynski, J., Temperature-induced optical phase shifts in fibers. *Applied Optics*, 1981, **22**:478–483.

16. Ramesh, S.K. and Wong, K.C., Design and fabrication of a fiber Bragg grating temperature sensor. *Proceedings of SPIE 3620*, 1999, 334–338.

17. Protopopov, V.N., Karpova, V.I., et al., Temperature sensor based on fiber Bragg grating. *Proceedings of SPIE 4083*, 2000, 224–228.

18. Zhan, Y., Xiang, S., et al., Fiber Bragg grating sensor for the measurement of elevated temperature. *Advanced Sensor Systems and Applications II*, 2005, 62–67.

19. Wang, Y., Tjin, S.C., and Sun, X., Measurement of temperature profile of concrete structures using embedded fiber Bragg grating sensors and thermocouples. *Proceedings of SPIE*, 2001, **4235**:288–297.

20. Huang, Y., Li, J., et al., Temperature compensation package for fiber Bragg gratings. *Microwave and Optical Technology Letters*, 2003, **39**(1):70–72.
21. Lin, Y.B., Chern, J.C., et al., The utilization of fiber Bragg grating sensors to monitor high performance concrete at elevated temperature. *Smart Materials and Structures*, 2004, **13**:784–790.
22. Zhan, Y., Cai, H., et al., Fiber Bragg grating temperature sensor with enhanced sensitivity. *Optical Fibers and Passive Components*, 2004, **5279**.
23. Hirayama, N. and Sano, Y., Fiber Bragg grating temperature sensor for practical use. *ISA Transactions*, 2000, **39**:169–173.
24. Kisala, P., Pawlik, E., et al., Fiber Bragg grating sensors for temperature measurement. *Lightguides and Their Applications II*, 2004, **5576**.
25. Wade, S.A., Grattan, K.T.V., et al., Incorporation of fiber-optic sensors in concrete specimens: Testing and evaluation. *IEEE Sensors Journal*, 2004, **4**(1):127–134.
26. Chuan Wang, Z.Z., Zhang, Z., and Ou, J., Early-age monitoring of cement structures using FBG sensors. *Smart Structures and Materials*, 2006, **6167**.
27. Kuang, K.S.C., Kenny, R., et al., Embedded fiber Bragg grating sensors in advanced composite materials. *Composites Science and Technology*, 2001, **61**:1379–1387.
28. Romero, M.A. Calligaris, A., Jr., and Camargo Silva, M.T., A fiber-optic Bragg grating temperature sensor for high voltage transmission lines. *IEEE Proceedings, Microwave and Optoelectronics Conference*, 1997, **1**:34–38.
29. Meltz, G., Morey, W.W., and Glenn, W.H., Formation of Bragg gratings in optical fibers by a transverse holographic method. *Optics Letters*, 1989, **14**(15):823–825.
30. Hill, K.O, Fiber Bragg grating technology fundamentals and overview. *Journal of Lightwave Technology*, 1997, **15**(8):1263–1276.
31. Albin, S., Enhanced temperature sensitivity using coated fiber Bragg grating. *Proceedings of SPIE*, 1998, **3483**.
32. Ong, K.C.G. and Akbarnezhad, A., Thermal stresses in microwave heating of concrete. Paper presented at Our World in Concrete and Structure, August 16–17, 2006, Singapore.
33. Bažant, Z.P. and Zi, G., Decontamination of radionuclides from concrete by microwave heating. I: Theory. *Journal of Engineering Mechanics*, 2003, **129**(7):777–784.

Chapter 6

Microwave heating cavities and applicators

6.1 INTRODUCTION

Three novel microwave-assisted methods relevant to the concrete industry for use in accelerated curing, selective demolition, and recycling of concrete were introduced in Chapters 2 to 4. The efficiency of these methods was illustrated using examples of results from experimental and numerical studies reported in the available literature. Although the microwave-assisted methods discussed seem to offer much promise in terms of efficiency, promoting widespread adoption is not feasible unless these are demonstrated to be efficient, economical, and easy to use without compromising safety and health issues during implementation. In this chapter, the basic configuration and the various components of a typical industrial microwave heating system suitable for such applications are introduced. After dealing with the general system configuration, focus is then placed on the design of the microwave applicators. The characteristics of the applicator can significantly affect the distribution and intensity of the incident microwave field and thus the effects of the microwave field on the material being processed. Therefore, familiarity with the design concepts and performance parameters related to microwave applicators is essential for the design of an efficient microwave heating system customised to suit the needs of a particular application.

6.2 THE MAIN COMPONENTS OF MICROWAVE HEATING SYSTEMS

The main components of a typical industrial microwave heating system for application in the concrete industry may be grouped into the following:

1. *Microwave generator unit*, which comprises active microwave components used to generate the microwave field needed.
2. *Power transmission unit*, which comprises a variety of waveguide components used to deliver the generated microwave power to the

material or load (e.g., concrete or recycled concrete aggregate [RCA]) being processed while minimising power reflection.

3. *Cooling unit*, which comprises pumps, pipelines, cooling tower, and heat exchanger to dissipate the internal heat generated during operation.

4. *Control unit*, which usually comprises a PLC (programmable logic controller) programmed to control microwave generation based on data input by the operator as well as the feedback from a network of sensors installed to regulate and control the various functions associated with microwave generation and delivery.

5. *Microwave applicator*, which is used to transfer the microwave power into the load while ensuring safety and health issues are in compliance.

In the following, the configuration and working principles of the above units and the role they play in the generation and delivery of microwave power are explained.

6.2.1 The microwave generator unit

The microwave generator unit is the heart of the microwave heating system. It consists of the microwave source and the power supply. The most commonly used source of microwave energy, primarily for reasons of efficiency, is the magnetron. Because of mass production, magnetrons at 2.45 GHz are particularly cheap; however, magnetrons are available for other frequency ranges as well. A typical 2.45-GHz magnetron is shown in Figure 6.1. Other sources available include traveling wave tube (TWT), klystron, gyrotron, and solid-state devices. Each has characteristics that can be exploited to suit the needs of the user.

As shown in Figure 6.1, the magnetron consists of a hot filament as the cathode (source of the electrons), which is placed in the center of an evacuated circular chamber. A permanent magnet is used to impose a magnetic field parallel to the filament to force the electrons, attracted to the outer part of the chamber, to spiral outward in a circular path rather than moving directly to the anode. As the electrons sweep past the cylindrical cavities spaced around the rim of the chamber, they induce a high-frequency electromagnetic field in the cavity. A portion of the field is extracted using an antenna and is directed to the waveguides through the magnetron's ceramic dome. To generate the microwave, the magnetron's filament should be kept at a high negative voltage. A high-power direct current (DC) power supply is used to provide such a high negative voltage. Figure 6.2 shows a 10-kW microwave DC power supply used commonly in industrial microwave systems. The output power of this power supply is adjusted through an analog signal, ranging from 1 to 10 VDC.

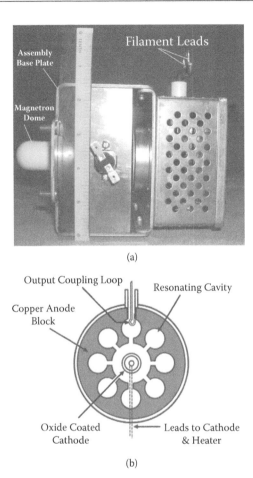

Figure 6.1 (a) A commercially available magnetron; (b) the internal structure of a magnetron.

To warm the magnetron's filament before commencing microwave generation, a filament transformer is used to preheat the magnetron. Figure 6.3 shows a typical filament transformer that can be used to heat the filament of a 2.45-GHz magnetron by passing a 30-amps current through it for almost 4 minutes.

6.2.2 Power transmission unit

The microwave power generated is delivered to the load (e.g., concrete or RCA) through a set of waveguide components. As shown in Figure 6.4, typical high-power configurations of the waveguide components usually

Figure 6.2 Switch-mode power supply.

Figure 6.3 A filament transformer.

include an isolator (circulator + water load), an impedance tuner, a power measurement tool, and some combinations of straight and bent waveguide sections. Depending on the particular application and cost considerations, some of these components may not be needed. In the following, the waveguide components used in a typical industrial microwave heating system are briefly described.

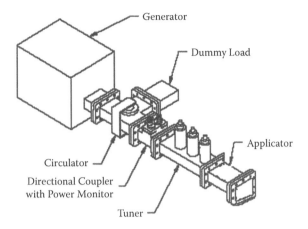

Figure 6.4 Configuration of a typical power delivery unit used in industrial microwave heating systems.

6.2.2.1 Isolator (circulator and water load)

As the power travels toward the load, as a result of reflections from surfaces of the passive microwave components (especially at bends) as well as the reflections from the surface of the load, a part of this power is reflected back toward the generation unit. The reflected power should be minimised through optimising the design of the microwave system and maximising the impedance matching (explained further in the chapter) to minimise reflections from the load surface. However, even the most sophisticated designs cannot completely prevent some amount of backward energy, and a percentage of the forward energy is usually reflected toward the microwave generation unit. If not dissipated before reaching the microwave generation unit, the reflected power could damage the magnetron and other active microwave components. The isolator is a two-port device made of a ferrite material and magnets that does not permit flow of the power in the reverse direction. For low-power applications, an isolator is normally used to protect the magnetron from the damaging effects of the reverse power. However, in high-power applications, instead of using the standard two-port isolator, an isolator is formed by connecting a circulator to the dummy waveguide load to absorb the reverse power. The circulator is a three-port device used to control the direction of power flow (Figure 6.5). Dummy loads are used to dissipate the microwave power with little or no reflection (Figure 6.5). The dummy loads used may be generally divided into wet loads or dry loads. Wet loads absorb the reflected microwave power by directing it into a high-loss fluid medium, usually water, whereas dry

(a)

(b)

Figure 6.5 Circulator (a) and water load (b).

loads absorb the reflected power by directing it into a high-loss solid. The high-loss solids used in dry loads are commonly made of silicon carbide. Both types of dummy load have distinct advantages and disadvantages. The most important advantage of wet loads is their small size and high-power rating as compared to dry loads [1]. Wet loads are often preferred in high-power microwave heating processes, whereas dry loads are preferred for applications that require standing waves to be minimised, including travel-ing wave applicators [1].

6.2.2.2 *Directional coupler and power monitor*

The microwave power actually generated and the power delivered to the load are usually different from the power estimated by the control systems based on the voltage provided and the designed efficiency of the active microwave components. This is mainly because of the differences between the actual efficiency and expected efficiency of the microwave generation system as well as the losses (because of energy dissipation or reflection) by the passive microwave components as microwaves travel from the power generation unit to the load. Measuring the actual forward power generated and power delivered to the load is important in identifying the energy loss incurred in the respective power generation and delivery units. In addition, measuring the reverse power reflected by the load or by the various passive components is important in the selection of a suitable isolator and in monitoring its performance to prevent any damage to the components of the microwave generation unit. The directional coupler is used to take sample readings of the power propagating in one direction, which are then used to estimate the total power propagating through the waveguide. The coupler then sends a signal, which is usually in the form of a DC voltage, to the meter or some other calibrated signal-measuring device. Similarly, dual directional couplers are used to take a sample of the power in both directions. Hence, they can be used to measure both the forward and the reverse power. Figure 6.6 shows a typical dual directional coupler and power-monitoring set.

6.2.2.3 *Tuner*

In microwave heating, the microwave power should be transferred from the generator to the load with minimal losses. The transfer efficiency of

Figure 6.6 Dual directional coupler and power-monitoring set.

microwave heating is limited by two factors, the energy absorbed by the walls of the transmission line through resistive heating and the power reflected from the load. In most cases, the power loss through resistive heating of the waveguide components is negligible, and the second factor plays a more dominant role. To achieve optimum power transfer from the source to the load, the load resistance (impedance) should be equal to the generator's internal resistance. This concept is known as *impedance matching.*

One common impedance-matching method is to insert a metallic element (stub) into the applicator (waveguide). By adjusting the position and the depth of insertion of the metallic element in the applicator, either the phase or the amplitude of the source impedance can be adjusted. Multistub tuners are normally used to adjust the position and depth of insertion via inserting one or more stubs into the applicator. When the change in the properties of the load during microwave heating is insignificant, manually adjusted tuners are usually used. The adjustment of the stubs through a manual tuner is mostly done based on the experience of the operator.

However, in applications where there is a significant change in the load impedance during microwave heating, manual tuning is rather inconvenient and too slow to keep pace with the changes. In this case, autotuners are usually used. Autotuners can achieve superior impedance matching by adjusting the stubs within a matter of milliseconds. Autotuners are usually more costly compared to manual tuners, but they have the potential to significantly increase the efficiency of microwave heating and increase the life of the microwave generator by minimising power reversals during processing. Considering the relatively high variations in the properties of concrete or RCA samples (the load) because of variations in their composition and water content, the use of autotuners in microwave systems for use in concrete industry applications is highly recommended. A typical industrial autotuner system is shown in Figure 6.7. The tuners should be located as close to the load being matched as possible [1].

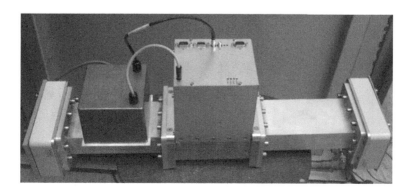

Figure 6.7 Autotuner.

6.2.2.4 Waveguides

A hollow metallic tube of either rectangular or circular cross section is generally used in practice to transfer microwave power from one component to the next or to the load. Such a structure is commonly known as a waveguide. Waveguides are usually made of aluminum, copper, or brass of various sizes. To choose the appropriate waveguide section, the operating microwave frequency, component availability, power rating, and costs should be considered. The standard waveguide sizes for 2.45-GHz frequency are listed in Table 6.1. The WR284 waveguide is the preferred choice of waveguide for systems operating at 2.45-GHz microwave frequency and for average power levels up to 6 kW. WR340 and WR430 are recommended for higher power levels. Figure 6.8 shows an aluminum WR430 waveguide.

Waveguide flanges are used to connect one waveguide section to another. Standard flange types are available for each waveguide type.

Table 6.1 Waveguides commonly used in industrial microwave heating at 2.45 GHz

Inside dimensions (in.)	Frequency band	Official designation		
		IEC	RCSC (ccc)	EIA (US)
2.84 × 1.34	s	R32	WG10	WR284
3.40 × 1.70	s	R26	WG9A	WR340
4.30 × 2.15	s	R22	WG8	WR430

Figure 6.8 Straight WR430 waveguide section with CR430 flange.

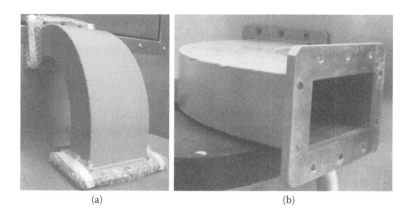

(a)   (b)

Figure 6.9 Waveguide bends: H bend (a) and E bend (b).

In most industrial heating applications, "flat-face" flanges are the most cost-effective choice, whereas "choke"-type flanges are more popular for military and communication applications. Most commercial waveguide components are made of aluminum because of its low cost and good performance. If heat loss is of major concern, copper waveguides may be more appropriate, although their costs usually outweigh the slightly lower heat loss advantage. Where sanitation is of high importance, such as in the food-processing industry, stainless steel waveguides may be used.

It is common in microwave transmission lines that the direction of transmission changes once or more times from the original direction of the launch waveguide. Waveguide bends are used to change the direction of the waveguide transmission system. H bend and E bend are two types of common waveguide bends used to change the direction of the transmission line in the horizontal and vertical directions, respectively (Figure 6.9).

An illustrative example showing how the various components of the microwave power transfer unit and microwave generation unit are configured for a typical industrial microwave heating system used in the separation of adhering mortar from RCA is shown in Figures 6.10 and 6.11. Figure 6.10 shows the magnetron being used to generate microwaves using the high voltage supplied at its filaments by the filament transformer. As can be seen, the system shown in Figure 6.10 uses a circulator and water load to dissipate the reflected power and thereby protect the magnetron. Figure 6.11 shows a typical transmission line, including a dual directional coupler to measure both the forward and the reverse power, a straight section rectangular waveguide, a waveguide bend to change the transmission direction, and an autotuner to maximise power absorption by the load placed in the multimode applicator (RCA processing chamber).

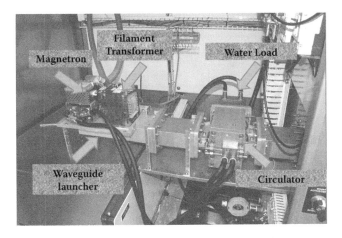

Figure 6.10 The magnetron, filament transformer, water load, and circulator.

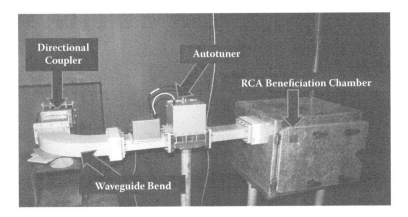

Figure 6.11 The directional coupler, autotuner, and multimode cavity used for separation of mortar from RCA.

6.2.3 Cooling unit

A water-cooling system is used to absorb the heat generated by the operation of the magnetron and the heat dissipated in the water load. The cooling system used in a typical industrial microwave heating system is usually divided into an internal closed cooling loop designed for water to circulate through the magnetron, circulator, and water load and an external cooling loop designed to dissipate the heat generated in the internal loop using a heat exchanger.

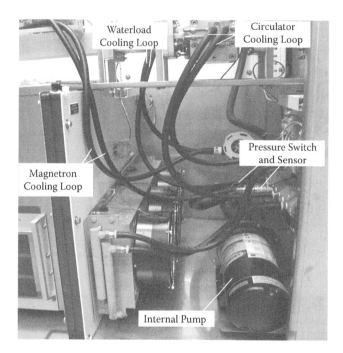

Figure 6.12 The internal cooling loop.

As shown in Figure 6.12, a typical internal cooling loop comprises a pump and three sets of closed pipeline loops. The first loop circulates water through the magnetron; the second and third loops provide the cooling water for the circulator and water load, respectively. The internal loop should be filled with distilled water to prevent the formation of deposits within the cooling passages. The presence of deposits in water can lead to overheating, which can shorten the life of the magnetron and circulator.

Figure 6.13 shows the components of a typical external cooling system that comprises a cooling tower, a pump, and a heat exchanger. The external cooling water absorbs the heat of the internal cooling water using the heat exchanger. The cooling tower is used to keep the temperature of the external cooling water within the prescribed range of 20°C–25°C. A pump is used to circulate the external cooling water through the heat exchanger and cooling tower.

6.2.4 Control unit

Control units are designed to ensure optimal and safe operation of the microwave heating system through a network of sensors, relays, and contactors controlled usually via a PLC programmed to operate the system

Figure 6.13 The external cooling loop comprising heat exchanger, cooling tower, and external pump.

according to input of the user and safety and health requirements. The various components of a typical control unit are briefly explained next.

6.2.4.1 Programmable logic control

The PLC is the heart of the control unit. A PLC is usually used to control the internal units of the microwave heating system and to communicate with external computers, processors, and Ethernet/Internet Protocol (IP) networks. The PLC includes three main modules: I/O (input/output) module, processor, and communication module. The I/O module is used to send and receive commands to and from the sensors and controllers. The output and input information, from the operator or the sensors, is processed within the processor unit according to the logic control program encoded in a programming language compatible with the PLC. The PLC communication module comprises Ethernet/IP and modem modules to communicate with the external computer and processors. The communication module can be used to change and modify logic programming and monitor the processor operation remotely. The various components of a PLC used to control a typical industrial microwave heating system are shown in Figure 6.14. PLCs are usually connected to a viewing panel through a local control module. The panel is used to receive the operator's input commands and show the system status prior to and during microwave generation.

6.2.4.2 Sensors and relays

Sensors and relays are used to monitor the operation of the various parts of the microwave heating system. A relay is an electronic switch that can stop

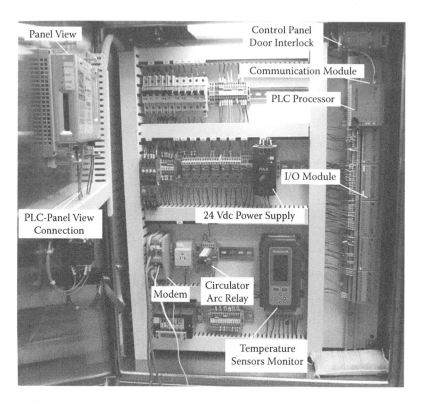

Figure 6.14 Control panel.

or start the operation of a specific part of the generator based on the sensors' signals. The most important sensors and relays used in typical industrial heating systems are as follows:

1. *Door Interlocks:* Both the generator cabinet door and the control panel door should be protected by safety interlocks to prevent generator operation when the door is opened or unlatched. These safety switches are monitored by the main control system to ensure operator safety and are checked for proper operation each time the system is started. If any door switch is cycled during system operation, the system is shut down, and a fault appears on the generator operator interface identifying the switch involved.

2. *Circuit Breaker Monitor:* This sensor is used to check whether the internal circuit breaker is switched on to provide power to the active microwave components.

3. *Circulator Arc Detector Relay:* An optical sensor is used to detect the arc in the circulator. This sensor is usually connected to a circulator

arc detector relay, which stops the generator's operation in case of arc detection.

4. *Internal Pump Pressure and Flow Sensors:* These sensors are used to ensure adequate water flow and pressure are provided for the cooling of the circulator, magnetron, and water load.

5. *E-Stop Relay:* Every microwave heating system should be equipped with an emergency stop (E-Stop). The E-Stop button should be located within easy reach in case of an emergency. In most state-of-the-art systems, the E-Stop function can also be activated remotely. The E-Stop relay stops generator operation on activation in an emergency.

6. *Coolant Temperature Sensor:* This sensor is used to control the temperature of the internal cooling water.

7. *Magnetron's Fan Sensor:* This sensor is used to check whether the magnetron fan is operating before starting microwave generation.

8. *External Cooling Water, Pressure, and Temperature Sensors:* These sensors should be located at the output of the heat exchanger to control the temperature and pressure of the external cooling water.

9. *Overload Relay:* This relay is used to stop microwave generation if the pump is not working properly.

10. *Magnetron Over-Current and Over-Voltage Relay:* These relays are used to stop generator operation in the event of detection of over-current and over-voltage conditions, which can damage the magnetron.

11. *Power Supply Sensors:* These sensors are used to monitor the operation status of the power supply and control preheating of the magnetron using the filament transformer.

6.2.5 Microwave applicator

Microwave applicators in various forms (waveguides, chambers, or cavities) are used to deliver the microwave energy to the load while ensuring the safety of operating personnel against microwave radiation. Because of the crucial role microwave applicators play in the microwave heating process, the remainder of this chapter provides an introduction to the microwave applicators, applicator design concepts, and the criteria for the selection of an optimal applicator type and configuration for the concrete industry applications discussed previously.

6.3 MICROWAVE APPLICATORS: INTRODUCTION AND DESIGN BASICS

This section provides a brief introduction to microwave applicators. The basics of operation and properties of the various types of microwave applicators suitable for use in microwave heating applications are discussed.

6.3.1 Applicators versus probes

The terms *applicator* and *probe* are commonly used in microwave engineering. Although the fundamental physics involved are similar, there are some practical differences in the applications and design considerations. These differences are usually related to the differences in the level of power, voltage, and currents in the particular applicator and probe. In microwave engineering, the term *applicator* refers to a device that applies microwave energy into a load at a level sufficient to temporarily or permanently change its material property or parameter. The change may involve a change in the temperature or moisture content, a chemical reaction catalysed by the microwave power, and so on. The term *probe* or *sensor* refers to a device used primarily to obtain information from the load being processed. For instance, a probe system may be used to apply a field onto a load to extract some information about the load's properties. An example of this is the probe used by the microwave dielectric measurement kit used to measure the dielectric properties of materials or loads as discussed in Chapter 2. The strength of the field transmitted by the probe is usually far lower than the level needed to cause any significant change in the properties/parameters of the material.

6.3.2 Far field versus near field

A near-field system is a system in which the distance between the applicator and the load is small compared to the wavelength. On the contrary, in a far-field system, the distance between the antenna (which is the term generally used instead of applicator/probe in near-field systems), and the receiver is usually large compared to the wavelength. In far-field systems, the electromagnetic field is in the form of travelling plane waves.

In near-field systems, the presence of the load (the material being heated) has a considerable impact on the field configuration. In other words, the load or material through its dielectric properties becomes a part of the system. However, in far-field systems, the effect of the load or material on the field's variables is often negligible. All the applicators used in the microwave-assisted processes discussed in this book are near-field.

6.3.3 Applicator performance parameters

6.3.3.1 Efficiency

The main components of a simplified microwave heating system are schematically shown in Figure 6.15. A source of microwave energy (e.g., a magnetron) feeds power to an applicator through a coupling system. The utility power P_u is supplied in the forms of alternating current (AC) or DC to the microwave source, which in turn supplies a power level P_s to the

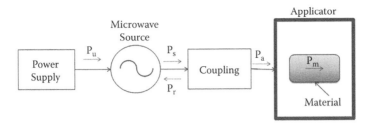

Microwave
Source

Figure 6.15 A simplified configuration of typical microwave heating systems. (From Mehdizadeh, M., *Microwave/RF Applicators and Probes for Material Heating, Sensing, and Plasma Generation: A Design Guide*. Norwich, NY: William Andrew, 2009. With permission.)

coupling system. A portion of the supplied power is reflected (P_r) because of mismatches; the rest is transmitted to the applicator (P_a). A portion of the applicator power is wasted (P_w); the rest is deposited into the load (P_m). The wasted power in most cases is caused by resistive heating in the metallic parts of the applicator. However, in some cases, other forms of power wastage may arise because of radiation or from the heating of other dielectric bodies present in the applicator.

The system efficiency of the microwave heating system is defined as the ratio between the power deposited into the load and the utility power. It is always desirable to maximise efficiency:

$$\eta_s = \frac{P_m}{P_u} \tag{6.1}$$

Similarly, the efficiency of the applicator together with the coupling system may be defined as the ratio between the power deposited into the load and the power supplied by the microwave source:

$$\eta_{ac} = \frac{P_m}{P_s} \tag{6.2}$$

The applicator efficiency can be similarly defined as the ratio between the microwave power deposited into the load and the power transmitted to the applicator:

$$\eta_a = \frac{P_m}{P_a} \tag{6.3}$$

The applicator efficiency is an important quality control measure in the design of applicators.

6.3.3.2 Fill factor

The fill factor of an applicator is a measure of the relative volume of the effective field region of the applicator, which is where the load is placed. The effective field region of an applicator is defined as the region over which the electromagnetic field is located. In other words, in regions outside the effective field region, the field is zero or of an insufficient amplitude for the intended function. The fill factor may be generally defined as

$$\varnothing = \frac{\iiint_{V_m} |I| dv}{\iiint_{V_t} |I| dv} \tag{6.4}$$

where I is the intensity of the electric field or magnetic field. The numerator is the field intensity integrated over the volume occupied by the load or material v_m. The denominator is the field intensity integrated over the volume of the effective field region. If the electromagnetic field in the applicator is nonuniform, Equation 6.4 tends to give a larger fill factor when the load is placed in regions with a stronger field intensity. If the electromagnetic field in the applicator is ideal, the equation can be simplified as

$$\varphi = \frac{V_m}{V_t} \tag{6.5}$$

This simplification can be a valid approximation in many practical cases. When designing an applicator, it is usually desirable to maximise the fill factor. This means that it is usually desirable to have a load that fills as much volume as possible and is placed at locations with the highest field intensity.

6.3.3.3 Field uniformity

Field uniformity is an important characteristic that varies considerably with the microwave frequency, design of the applicator, and dielectric properties of the load or material. Maximising the field uniformity in the applicator is desirable in most applications. In concrete technology, irrespective of whether the application requires achieving uniform heating or nonuniform surface heating of the load (concrete/RCA), the uniformity of the field intensity inside the applicator is usually desirable to ensure uniformity in the quality of the concrete/RCA processed between batches. Therefore, availability of a means to measure uniformity of the field inside an applicator is highly important. Field uniformity can be calculated using the following equation:

$$\rho = \frac{\left|I_{max}\right| - \left|I_{min}\right|}{\left|I_{max}\right| + \left|I_{min}\right|} \times 100\% \tag{6.6}$$

where $\left|I_{max}\right|$ and $\left|I_{min}\right|$ are the maximum and minimum amplitudes of the electric or magnetic field intensity, depending on the particular application.

To control the uniformity of the electromagnetic field in a microwave applicator, it is essential to be familiar with the possible causes of non-uniformity. One of the main causes of nonuniformity is the occurrence of standing waves because of the well-known wavelength effect that happens in large applicators (large compared to the wavelength). A standing wave is a superannuation of a reflected wave from the surface of a conductive plane and a forward wave. In addition, another important parameter affecting field uniformity is the load or material load being processed. The load tends to change the electromagnetic boundary conditions of the field and leads to some nonuniformity. Materials with higher dielectric losses or conductivity tend to cause more nonuniformity in the field.

6.3.4 Resonance

Resonance is important in microwave heating applicators and is commonly used to influence the intensity of the interactions between the field and the load or material. Applicators may be designed as resonant or nonresonant. Resonant applicators are usually enclosed metallic structures with dimensions typically similar to or larger than the wavelength of the operating frequency. With this definition, practical resonant applicators operate exclusively at microwave frequencies of 300 MHz to 30 GHz, which correspond to the wavelengths of 1 cm to 1 m. In systems operating at industrial, scientific, and medical (ISM) frequencies, the impact of resonance is mainly in terms of the field configuration inside the applicator. Resonance tends to increase the intensity of the field. Therefore, the absence of resonance means the electric field is not at its strongest state. Resonance can become unnecessary in some situations, including heating of materials with very high dielectric loss. This is because, as shown by Equation 1.33, there is less need for a high electric field if the material has a high loss factor. However, despite its advantages, resonance may give rise to complexities in hardware design and in the operation of the system because of tuning difficulties and the like. On the other hand, nonresonant applicators have advantages, including having a more simplified structure and the lack of a need for tuning. Therefore, it is common and desirable to avoid resonance when possible. Transmission line or traveling wave applicators are among the more commonly used types of nonresonant applicators (Figure 6.16). In such applicators, the material is placed in the path of the travelling waves

Figure 6.16 (a) Cross section of a strip transmission line; (b) cross section of a rectangular waveguide transmission line.

in a transmission medium, and the electric field present in the applicator interacts with the load or material.

6.3.6 Material loading effect

The placement of a material in an applicator affects considerably the field configuration because the properties and geometry of the load or material become a part of the electrical properties of the system. This makes the design of the applicators particularly challenging, especially when properties of the load or material change from sample to sample, with time, with load temperature, and so on.

The impact on the field configuration is mainly caused by changes in the electromagnetic boundary conditions of the applicator when the load or material is added to the system. Because of such changes in the boundary conditions, the simplified analytical formulas for the applicators before being loaded become inaccurate, and in most cases finite element analysis is required to solve Maxwell's equations when the new prevailing boundary conditions are invoked.

Another important challenge in designing microwave applicators is the nonlinearity caused by the change in the load or material properties as the load's temperature increases as heating proceeds. Because of the evaporation of water from free surfaces of the load or material, its water content and thus dielectric properties change during the microwave heating process. The change in the material properties results in changes that affect the design parameters of the applicator, including resonant frequency, quality factor, and so on.

6.4 APPLICATORS USED IN HEATING SYSTEMS

In this section, the types of applicator most commonly used in microwave heating applications, especially those described in the previous chapters of this book, are introduced.

6.4.1 Multimode microwave cavities

Multimode resonant cavities are the most commonly used type of applicator in microwave heating systems. Because of their robustness and simplic-
ity of design and manufacture, multimode resonant cavities are used as
applicators for domestic microwave ovens. In principle, any metallic enclosure with dimensions a few times larger than the microwave wavelength
may serve as a basic multimode applicator for microwave heating purposes.
In such cases, the microwave energy is transmitted using a transmission
line or waveguide to the cavity, which contains a lossy material. The ease
of operation and elimination of the need for regular tuning are among the
main advantages of such basic multimode applicators. Unlike other types
of applicators available that can be used in a variety of applications, multimode cavities are used only in microwave heating and plasma activation
applications. The multimode applicators are considered the most suitable
type of applicator for microwave-assisted separation of mortar from RCA
and microwave-assisted accelerated curing of precast concrete. Therefore,
the basics for the design of such applicators are discussed in detail in the
following section.

Multimode microwave cavities can be made in different three-dimensional
(3D) configurations with dimensions that are at least two times larger than
the operating microwave wavelength. Rectangular microwave cavities are
preferred because they are both simple to analyse and manufacture. The
exact analytical solutions for various basic shape multimode cavities are
available in the literature [2–5]. A typical microwave cavity with an arbi-
trarily selected set of dimensions is schematically shown in Figure 6.17a [6].

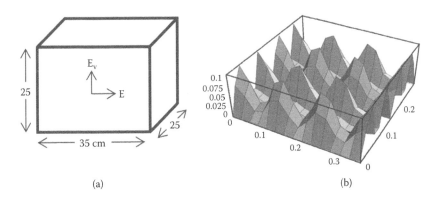

(a)   (b)

Figure 6.17 Power density distribution under a TE_{324} mode in a rectangular micro-
wave cavity. (From Mehdizadeh, M., *Microwave/RF Applicators and Probes for
Material Heating, Sensing, and Plasma Generation: A Design Guide.* Norwich,
NY: William Andrew, 2009. With permission.)

The cavity is assumed to operate at 2.45 GHz and TE_{324} mode. The TE_{324} is a transverse electric (TE) mode, which by definition has no electric field in the z direction and has 3, 2, and 5 variations in the x, y, and z directions, respectively. As shown in Chapter 1, the energy density in an electromagnetic field is proportional to the square of the electric field intensity norm. Therefore, it can be assumed that the energy density in the multimode cavity shown in Figure 6.17a is proportional to the norm of the intensities of the x and y electric field components, that is, $\left|E_x^2 + E_y^2\right|$. By taking this into account, a numerical simulation of the microwave field shows that the normalised power density at the observation plane located in parallel to the x-z surface in the center of the cavity follows the 3D plot shown in Figure 6.17b [6]. As shown, there are three and four obvious variations along the x and z directions, respectively. This indicates poor power uniformity within the cavity as the power density can vary from a maximum value to a minimum of zero every few centimetres. Using Equation 6.6, the field uniformity can be estimated as 100%, which is the worst-possible uniformity for an applicator. However, as discussed previously, the nonuniformity of a microwave field is undesired in most heating applications. Thus, maximising the field uniformity is one of the main objectives in the design of multimode cavities. In the following, various techniques for improving the field uniformity in multimode cavities are discussed.

6.4.1.1 Resonant frequency and possible modes

Identifying the relationship between the excited mode, resonant frequency, and cavity dimensions is crucial in designing efficient multimode cavities. In general, such a relationship should be determined by finding the eigenvalues of the general wave equations in the phasor form, expressed as follows:

$$\nabla^2 E + k^2 E = 0 \qquad (6.7)$$

$$\nabla^2 H + k^2 H = 0 \qquad (6.8)$$

where ∇ is the Laplacian operator, and k is the wave number, which has the following relationship with the angular frequency ($\omega = 2\pi f$) in radians/second, permittivity ε in farads/metre (F/m), and magnetic permeability μ in henries/metre (H/m).

$$k = \omega\sqrt{\mu\varepsilon} \qquad (6.9)$$

When considering an empty rectangular cavity, the following analytical solution for the relationship between mode number, cavity dimensions, and resonant frequency can be easily obtained [4]:

$$f_0 = \frac{C}{2} \sqrt{\left(\frac{m}{a}\right)^2 + \left(\frac{n}{b}\right)^2 + \left(\frac{k}{l}\right)^2} \qquad (6.10)$$

where f_0 is the resonant frequency in hertz (Hz) for the TE_{mnk} or TM_{mnk} mode in a rectangular multimode microwave cavity with length, width, and height of a, b, and l (in metres), respectively, and C is the velocity of light. The indices of the TE or TM (transverse magnetic) mode (i.e., m, n, and k) can be any integer, including zero. However, as a general rule, there cannot be any mode with two zero indices, and if one of the indices is zero, the mode can be either TM or TE but not both. When none of the indices is zero, both TE and TM modes can exist at the same resonant frequency.

Equation 6.10 implies that a cavity with known dimensions resonates at only certain distinct frequencies. In fact, this is a general rule for multimode cavities, regardless of their dimensions and the presence of the load. The magnetron used in the majority of heating applications can operate usually at only one ISM frequency with a relatively limited bandwidth. For instance, the 2.45-GHz magnetrons commonly used in domestic microwave ovens usually have a bandwidth of about 50 MHz over which the output energy is spread. The analytically calculated number of possible resonant modes for the bandwidth of 50 around the resonant frequency of 2.45 GHz for a rectangular cavity of dimensions 25 × 30 × 35 cm is shown in Figure 6.18. The variation in the number of possible modes with an increase in the size of the cavity is also illustrated. As shown, the number of possible modes increases with an increase in the size of the cavity. Figure 6.18 shows that doubling each dimension of the cavity may lead to an increase in the number of possible modes from 6 to 40. The existence of additional modes would improve the uniformity of the field in the cavity compared to when only one mode is present, as shown in Figure 6.17. This is because of superposition of the field from various modes, which reduces the probability of the existence of true field nulls. With this in mind and taking into account the increase in the possibility of having a higher number of modes with an increase in the size of the cavity, larger cavities are more likely to lead to a more uniform energy deposition.

Identifying the relationship between resonant frequencies, number of modes, and cavity dimension is not usually as simple as the demonstrated case of the simple empty rectangular cavity. For cavities with more complex shapes and material-loaded cavities, the concept of specific indices is no longer justified, and analytical solutions are not available. Fortunately, for cavities used as applicators in microwave heating systems, identifying the exact microwave modes in the cavity is not essential as a part of the design process, and such analyses are performed usually for academic interest to understand the operation of such multimode cavities. In such cases,

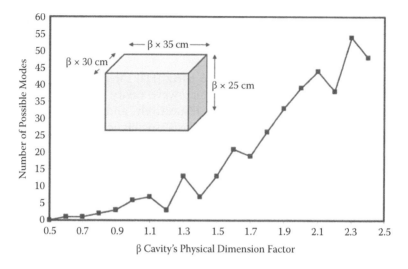

Figure 6.18 Variation in the number of possible resonant modes with the size of a cavity for a rectangular cavity at a microwave frequency of 2.45 GHz. (From Mehdizadeh, M., *Microwave/RF Applicators and Probes for Material Heating, Sensing, and Plasma Generation: A Design Guide.* Norwich, NY: William Andrew, 2009. With permission.)

cavities are usually studied using empirical probing measurement methods and numerical methods [7]. A variety of commercial numerical simulation packages based on finite-element or finite-difference time domain methods are available to perform cavity studies to investigate the performance of alternative cavity designs for a certain application. One of the major capabilities of state-of-the-art electromagnetic field analysis software such as COMSOL is their multiphysics capability, which allows coupling of various analysis, including electromagnetic, thermodynamic, and structural analyses to study the behaviour of particular materials subjected to microwaves in a multimode cavity [8,9]. Various examples of modeling performed using such software packages were presented in Chapters 2 to 4 to illustrate the capabilities of the various microwave-assisted methods introduced.

6.4.1.2 Improving the heating uniformity in multimode cavities

As discussed, one of the major challenges in designing multimode cavities for material processing is improving the field uniformity in the cavity. The variations in the field intensity because of the modal configuration lead to temperature nonuniformity in the load placed in the cavity. However, besides the modal configuration, the temperature distribution in the load heated in a multimode cavity is also affected by a number of other parameters. One

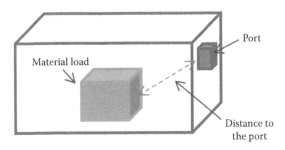

Figure 6.19 Proximity of the load material to the feed point. The surface closer to the feed point tends to heat faster.

of the major parameters is the penetration depth of fields in the lossy load material. As discussed in Chapter 1, the penetration depth of a plane wave in a lossy material is inversely proportional to the attenuation factor of the material and can be estimated using Equation 1.23. Regardless of the field distribution, the penetration depth of microwaves in a highly lossy material such as concrete can be a major factor affecting the heating uniformity in multimode cavities. Another factor that could possibly affect temperature uniformity in multimode cavities is the geometry of the material. The addition of the load into a multimode cavity considerably changes the field distribution and sets new boundary conditions for the field. The shape of the load also considerably affects the configuration of the fields inside the load material, especially when the relative permittivity is high. The edges of the objects with complex shapes usually receive more intense field and are therefore heated faster. For instance, the narrow end of a triangular-shaped load material is heated faster than its wider end. The third factor affecting the heating uniformity in multimode cavities is the proximity of load material to the feed point (Figure 6.19). The side of the load closer to the feed point usually receives more energy and is heated faster than sides located farther away. The nonuniformity of heating caused by this factor increases when the load material is placed closer to the feed point. Various methods are usually applied to reduce the effect of load proximity to the feed point. These include the use of multiple ports and rotation of the load. The latter method is commonly used in domestic microwave ovens and has proved to be relatively effective. The first method also has the advantage of exciting more modes, which could lead to better uniformity.

The microwave field uniformity in multimode cavities can be improved considerably through basic cavity design. By taking into account the causes of nonuniformity mentioned on top of the field modal configuration, a number of approaches may be adopted to improve uniformity of the field in multimode cavities. The first approach for improving the uniformity of heating in multimode microwave cavities is by increasing the number of excited

Figure 6.20 A large hexagonal-shape microwave cavity to achieve uniform heating at 2.45 GHz. (Courtesy of Vötsch Industrietechnik GmbH, Produktbereich Wärmetechnik, Greizer Str. 41-49, 35447 Reiskirchen-Lindenstruth, Germany.)

modes. The simplest method for increasing the number of excited modes is to increase the size of the multimode cavity with respect to the wavelength of the operating microwave frequency. An example of a relatively large (180 × 155 × 330 cm) microwave cavity with 962 excited modes at 2.45 GHz is shown in Figure 6.20 [10]. Such a cavity can achieve considerably better field uniformity compared to conventional domestic microwave ovens, which typically have from four to six excited modes.

Aside from using larger cavities, another method to increase the number of excited modes in multimode microwave cavities is to use angular or curved walls. In cavities with parallel walls, such as rectangular cavities, the reflection of microwave beams from parallel walls leads to the formation of multiple parallel beams, which tend to excite the same modes. The use of curved or angular walls can lead to multiple reflected beams, each exciting a different mode. This concept has been applied in the example cavity shown in Figure 6.20 through the use of a hexagonal shape rather than a simple rectangular shape.

The second approach for improving heating uniformity in multimode microwave cavities is by mechanically moving the load in the cavity. For example, the use of a turntable is an inexpensive and relatively effective method commonly used in domestic microwave ovens to improve heating uniformity. The load placed on the turntable continuously stirs the field by modifying the boundary conditions as it revolves within the cavity. This effect is more prominent for loads with less-regular shapes. However, it should be noted that, although the use of a turntable improves heating uniformity compared to the static case, it does not lead to overall uniformity. Circular symmetrical nonuniform heating is commonly observed in most cavities with a turntable. The pattern of this nonuniformity depends on the field pattern of the material-loaded cavity.

Another mechanical tool with a relatively similar function as the turntable that is commonly used to improve heating uniformity in microwave ovens is the mode stirrer (Figure 6.21). A mode stirrer is a metallic device that disturbs the field by moving continuously in the multimode cavity. The periodic changes in the field boundary conditions as a result of the disturbance caused by the stirrer tend to change the location of the true nodes and nulls in the cavity and thereby improve heating uniformity.

The effectiveness of mode stirrers varies with the type of load material as well as the design and location of the mode stirrer. It has been shown that when thick and highly lossy loads are placed in the cavity, the impact of

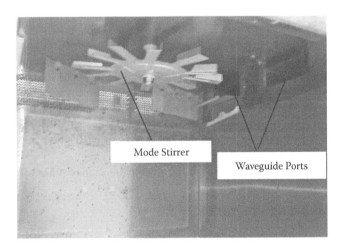

Figure 6.21 A mode stirrer used to improve heating uniformity in a domestic microwave oven.

mode stirrers tends to be minimal [11]. However, mode stirrers can considerably improve heating uniformity when small and low-loss material loads are heated. The location of the mode stirrer also affects its effectiveness. Placing the mode stirrer close to the microwave feed point can improve efficiency [12].

The third general approach to improve heating uniformity in a multimode cavity is by using higher frequencies. In a multimode cavity with fixed dimensions, heating uniformity increases generally with an increase in the microwave frequency. This is mainly because a larger number of wavelengths can be fitted in a multimode cavity with a particular fixed dimension, so a larger number of operating modes can exist at higher frequencies than at lower frequencies. In addition, at higher frequencies, the distance between the nodes and antinodes of standing waves decreases with an increase in microwave frequency, leading to improved conventional heat transfer between adjacent nodes and nulls. Aside from conventional 2.45-GHz magnetrons, those of higher frequencies, including 5.8-GHz magnetrons are readily available and are gaining popularity [13]. Considerably higher frequencies, including 24 GHz generated by gyrotrons, have also been used in applications such as ceramic sintering.

The fourth method used to improve heating uniformity in multimode cavities is called *frequency sweeping*. The working principle of this method is based on the fact that a larger number of modes may be excited in wider bandwidths. The presence of a larger number of modes in turn leads to a better heating uniformity. However, it is well known that achieving a wide bandwidth around a fixed frequency at high powers is challenging. One innovative alternative method for creating broad bandwidths is by high-speed frequency sweeping, which may create a relatively uniform time-averaged power density in the multimode microwave cavity. A variety of microwave electron devices, including helix TWTs, can be used to achieve the desired frequency sweeping to cover a broad bandwidth, such as 2–8 GHz. However, the variable-frequency microwave generators are highly expensive and are therefore not commonly used in normal heating applications. An alternative method to improve field uniformity is to use two or more magnetrons with different frequencies instead of a single variable magnetron. This method can be used to excite a large number of modes at considerably less cost than variable-frequency magnetrons.

6.4.2 Open-ended applicators

The multimode cavities discussed in the previous section are suitable applicators for applications such as microwave separation of mortar from RCA and microwave-assisted accelerated curing of concrete, for which the load shape and configuration allow it to be surrounded by the applicator and be uniformly heated from all the faces of the load as desired. However, there are applications, such as microwave-assisted selective demolition of concrete and hyperthermia or diathermy, for which only a certain part of the large load is to be exposed to microwaves. As shown in Chapter 3, in such applications, usually an open-ended waveguide is used to apply microwave power directly to the surface of the load. In the following, the basic design concepts of open-ended applicators are reviewed. The focus is placed on applicators suitable for use in microwave-assisted selective demolition of concrete.

Figure 6.22 shows a simple schematic representation of an open-ended applicator. The fringing fields leaving the open end of the applicator interact with the load being heated. The fringing field is a modified field that usually occurs in open-ended applicators. When the open-ended applicator is used to heat a load, the reflected power should be minimised to improve the efficiency of the process. Open-ended waveguide applicators usually have limited field coverage, which is confined to the applicator opening in the x and y directions and the penetration depth of microwave in the load in the z (propagation) direction. As shown in Chapter 3, the penetration depth of a microwave varies with the dielectric properties of the material and is usually characterised by the attenuation factor. In the case of concrete, the penetration depth and thus the field coverage can vary significantly with variations in the water content of the concrete.

The open-ended applicators can be generally divided into three groups: transverse electromagnetic (TEM), TE, and TM. The working principle of the applicator differs considerably depending on whether it is a TEM or TE/TM applicator. In TEM applicators, the electric field is perpendicular to the cross section of the waveguide. This requires the presence of at least two conductors surrounded by an open space or by dielectrics. Examples of TEM applicators include coaxial waveguides (Figure 6.23a) or

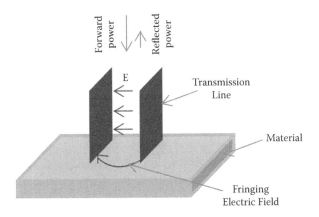

Figure 6.22 Schematic representation of an open-ended applicator.

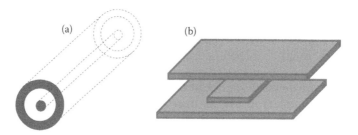

Figure 6.23 (a) Coaxial applicator; (b) strip transmission line.

strip transmission lines (Figure 6.23b). TEM applicators may operate over a wide frequency range without any lower bound. The upper bound of the operating frequency range of TEM applicators occurs when the wavelength reaches or is close to the dimensions of the applicator cross section.

Contrary to TEM applicators, the direction of the field in TE or TM applicators is not limited to the cross section of the applicator, and electric and magnetic field components are also present in the direction of propagation. Unlike TEM applicators, which usually require at least two conductors, TE and TM applicators can consist of only one conductor, such as a hollow waveguide. In addition, unlike TEM applicators, there is a lower limit for frequency at which TE and TM waveguides can operate. This frequency is referred to as the cutoff frequency. The most popular examples of TE and TM applicators are open-ended waveguides and waveguide horns (Figure 6.24).

The following reviews some of the basic working principles of open-ended coaxial applicators with focus on their application in microwave drilling and open-ended hollow rectangular waveguides with focus on their application in microwave-assisted selective demolition of concrete discussed in Chapter 3.

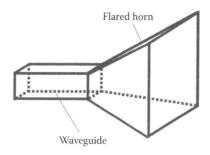

Figure 6.24 Flared horn as an open-ended microwave applicator.

6.4.2.1 Open-ended coaxial applicator

In Chapter 3, microwave drilling of concrete developed by Jerby and Dikhtyar was introduced as a novel application of microwave heating in concrete technology [14]. In a microwave drill, a coaxial applicator acting as a concentrator is used to concentrate the microwave energy at a very small spot, normally much smaller than the microwave wavelength itself. The coaxial applicator used in a microwave drill consists of a movable center electrode that concentrates the microwave power at a specific point. The concentrator itself is then inserted into the molten hot spot to create a crater (Figure 6.25).

Apart from application in microwave drilling of concrete, coaxial applicators are also commonly used as probes in dielectric measurement kits used to determine the dielectric properties of materials. Coaxial applicators can be used at a broad frequency range from kilohertz to tens of gigahertz. This is considered one of the main advantages of coaxial applicators over open-ended waveguide applicators, in which the operating frequency is highly affected by the dimensions of the waveguide. The field configuration in coaxial waveguides is schematically shown in Figure 6.26.

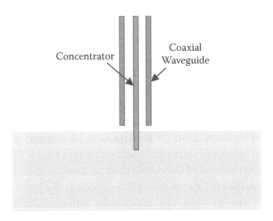

Figure 6.25 Schematic representation of microwave-drilling apparatus.

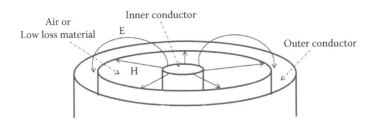

Figure 6.26 Field configuration in an open-ended coaxial applicator.

One of the main objectives in the design of coaxial applicators is to minimise the losses incurred. This is especially important when coaxial waveguides are used in measurement and monitoring applications, especially when a material with low dielectric loss is monitored. Based on the results of a series of comprehensive studies on the effects of various parameters on applicator losses, the following equation has been proposed to estimate coaxial applicator losses:

$$\alpha = \frac{R_s \sqrt{\varepsilon_r'}}{2b\eta} \left[\frac{1 + \dfrac{b}{a}}{\ln\left(\dfrac{b}{a}\right)} \right] + \frac{\pi f \varepsilon_r''}{C \sqrt{\varepsilon_r'}} \qquad (6.11)$$

where α is the total applicator losses (nepers/m), R_s is the surface resistivity of the metal, η is the impedance of free space (= 377 ohms), C is the speed of light, and a and b are, respectively, the inner and outer radii of the applicator. ε_r' and ε_r'' are the dielectric constant and dielectric loss of the applicators' dielectric, respectively. The first and second terms of Equation 6.11 represent, respectively, the metal losses and the dielectric losses in the dielectric between the inner and outer conductors. The first term of the equation shows that the metal losses can be reduced by increasing the radius of the coaxial applicator. This suggests that when using coaxial applicators to measure dielectric properties, the availability of samples large enough to accommodate larger coaxial applicators may lead to increased measurement precision by reducing the applicator losses. In addition, the results of mathematical optimisation of Equation 6.11 suggest that, for coaxial applicators filled with polytetrafluoroethylene (PTFE), a b/a ratio of 3.6 is the optimal configuration of the applicator to minimise metal losses, which are the most significant losses observed in coaxial applicators. Similarly, the optimal b/a ratio for coaxial sensors with different dielectric fillings can be estimated using Equation 6.11.

Equation 6.11 also shows that the dielectric loss in the coaxial applicator, which is represented by the second term in the equation, can be minimised by using low-loss dielectrics. The ideal solution is of course to use no dielectric at all, if possible.

6.4.2.2 Open-ended waveguide and horn applicators

One major difference between open-ended waveguide applicators and coaxial applicators is the size of the area affected by the applicator or the applicator coverage volume. The applicator coverage in coaxial applicators is usually smaller than the wavelength of the operating frequency, whereas in waveguide applicators, the coverage volume is of comparable size or larger

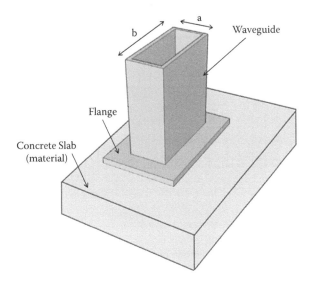

Figure 6.27 Configuration of an open-ended waveguide applicator.

than the wavelength of the operating frequency. The larger coverage area of the waveguide applicators may result in a higher processing rate because of the possibility of treating a larger area at one time. The most common open-ended waveguide applicator used is the rectangular waveguide in the TE_{10} mode, although circular waveguides can also be used. Figure 6.27 shows the arrangement of a rectangular open-ended waveguide applicator. TE_{10} is the most commonly excited mode in most waveguide applicators, and most commercial waveguides are designed for use in this mode. The electric and magnetic field configuration in TE_{10} in a rectangular waveguide is shown in Figure 6.28.

Aperture admittance of open-ended waveguide is a crucial indicator for the prediction of the reflection performance for use in impedance matching and thus is an important design parameter. Admittance Y is an indicator of how easily a device allows a current to flow and is the inverse of impedance Z, which is a measure of the resistance of the device to a current. The following equation has been derived for rectangular waveguide applicators [15]:

$$Y = G + jB$$

$$= \frac{2j}{\pi abk_1} \int_0^a \int_0^b (b-x) \left[K_2(a-y)\cos\frac{\pi y}{a} + \frac{a}{\pi} K_1 \sin\frac{\pi y}{a} \right] G_e d_y d_x \tag{6.12}$$

where G is the real part of admittance, which is referred to as conductance; B is the imaginary part of admittance, which is referred to as susceptance;

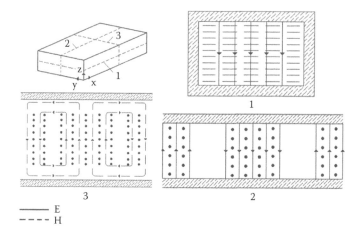

Figure 6.28 Field lines for the TE_{10} mode in a rectangular waveguide.

a and b are the width and height of the waveguide, respectively; and k_1 is the dominant mode number that can be determined using the following expression:

$$k_1 = \sqrt{k_0^2 - \left(\frac{\pi}{a}\right)^2}$$

(6.13)

where k_0 is the wave number in the free space, which can be determined using the following equation using the wavelength in the free space λ_0:

$$k_0 = \frac{2\pi}{\lambda_0}$$

(6.14)

K_1, K_2, and G_e parameters in Equation 6.12 can be determined using the following equations:

$$K_1 = k_r^2 + \left(\frac{\pi}{a}\right)^2$$

(6.15)

$$K_2 = k_r^2 - \left(\frac{\pi}{a}\right)^2$$

(6.16)

$$G_e = \frac{1}{r} e^{-jk_r\sqrt{x^2+y^2}}$$

(6.17)

The k_r parameter in Equation 6.17 can be obtained using the following expression:

$$k_r = k_0 \sqrt{\varepsilon_r} \tag{6.18}$$

The solution to Equation 6.12 can be obtained using various electromagnetic simulation packages such as COMSOL or mathematical simulation software such as Mathematica and MATLAB®. The distribution of the electric field in a rectangular open-ended waveguide is shown in Figure 6.29.

One type of open-ended waveguide applicator is the flared horn open-ended applicator, in which the side of the applicator closest to the load is flared to provide an incident area that is larger than the waveguide cross section. Horn open-ended applicators are commonly used in a variety of applications, including sensing and heating applications. A rectangular waveguide horn is shown in Figure 6.24. One advantage of open-ended horns over open-ended waveguide applicators is the higher penetration of the fringing field into the load. In open-ended horns, the penetration depth of the field is usually limited only by the load's penetration depth, which is determined by the material's attenuation factor as described in Chapter 1. The higher penetration depth of horn applicators is mainly because the

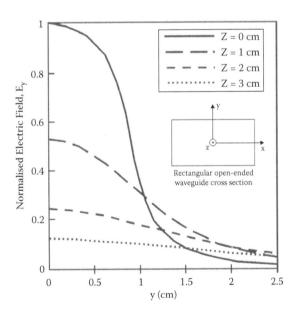

Figure 6.29 Electric field distribution in a rectangular waveguide applicator in TE_{10} mode. (Reprinted, with permission, from Habash, R.W.Y., Non-invasive Microwave Hyperthermia, PhD dissertation, Indian Institute of Science, Bangalore.)

single-mode transmission in the waveguide is usually transformed into multimodal propagation in the horn. This is the case; within a typical open-ended waveguide, the penetration of the field is usually controlled by the dimensions of the waveguide. The smaller the waveguide cross section, the smaller the depth of penetration into the load will be. Horn applicators can be used as an easy means to increase the coverage area of the microwave applicators in applications such as microwave-assisted selective demolition of concrete and microwave-assisted removal of tile and surface finishes from concrete floors and walls as discussed in the previous chapters.

6.4.3 Continuous-flow applicators

A number of applications of microwave heating in concrete technology, including accelerated microwave curing of prefabricated concrete components and microwave-assisted separation of mortar from RCA, can take advantage of continuous-flow processing, which leads to lower labour costs and higher outputs compared to batch processing. The principles of microwave heating and applicator design discussed previously apply also to continuous flow applicators and processes. However, there are special considerations that should be taken into consideration in the design and planning of continuous microwave heating processing. In continuous-flow applicators, the microwave power is supplied to a cavity with microwave-sealed openings in the form of open-ended ports. The load or material is conveyed via conveyor belts through the cavity via the input port before exiting via the output port. Providing a microwave attenuation mechanism at each port of such continuous-throughput cavities is essential to ensure the safety of operators in the presence of the high-power microwave field. Figure 6.30 shows a typical continuous-throughput microwave heating system in which the load is conveyed through the microwave cavity at a constant velocity via a conveyor belt system.

When processing materials using a continuous-throughput microwave heating system, it is crucial to estimate the power required to process the load volume being fed into the microwave system by considering the velocity of the conveyor belt. The properties and size of the components treated by such continuous microwave systems could vary considerably depending on the particular application, including accelerated microwave curing of precast concrete. The selection of an inappropriate power level or lack of a dynamic mechanism for feedback control of the microwave power based on the variations in the volume and properties of the load during processing could result in nonuniformity in the quality of the microwave-treated loads. The total microwave power needed in a continuous-throughput microwave heating system can be estimated using the following equation:

$$P = R_{pr}C_p\Delta T + h_{fg}R_{sol} \tag{6.19}$$

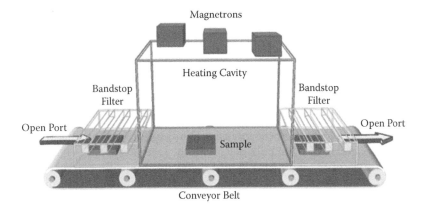

Figure 6.30 Simplified schematic representation of a typical continuous-throughput microwave heating system. (Reproduced by permission of IOP Publishing, from Clemente-Fernández, F.J., Monzó-Cabrera, J., et al., A new sensor-based self-configurable bandstop filter for reducing the energy leakage in industrial microwave ovens. *Measurement Science and Technology*, 2012, **23**(6):065101. doi:10.1088/0957-0233/23/6/065101.)

where P is the total power required (in kilowatts); C_p is the specific heat of the load (in kJ/kg °C); h_{fg} is the evaporation heat for the water present in the load, which is typically equal to 2257 kJ/kg at 100°C; R_{sol} is the rate of evaporation; and ΔT is the temperature rise in degrees centigrade. It should be noted that the required microwave power estimated using Equation 6.19 does not account for heat losses. In addition, the equation assumes that microwave heating is the only source of heating energy in the continuous-throughput cavity. This might not always be the case as additional sources of heat energy such as hot gases are used commonly in many commercial industrial microwave heating systems to preheat the load. In such cases, the equation should be modified to account for the heat energy generated by other supplementary sources of energy.

Alternatively, a more precise method to estimate the microwave power required by such continuous-throughput microwave heating systems to achieve heating at a particular temperature can be estimated by considering the relationship between the field intensity, microwave frequency, and volume and dielectric properties of the load. The microwave power dissipated in a load of volume V can be estimated using the following equation:

$$P = Kf\varepsilon_r''E^2V \tag{6.20}$$

where $K = 5.56 \times 10^{-11}$, E is the electric field intensity in the load (V/m), f is the microwave frequency (Hz), and ε_r'' is the loss factor of the load. By combining this equation with the heat transfer equation, the temperature of the load as a function of time $T(t)$ can be estimated using the following formula derived by Boldor et al. [16].

$$Kf\varepsilon_r''E(t)^2 = -\lambda\rho_0\frac{dA}{dt} + C_p(t)\rho_0\left[1 + A(t)\right]\frac{dT}{dt} \tag{6.21}$$

Here, $E(t)$ is the electric field intensity (V/m) in the load at time t, t is the time (in seconds), ρ_0 is the bulk density of the load (kg/m^3), C_p is the specific heat of the load at time t (in kJ/kg °C), and $A(t)$ is the moisture content of the load at time t. To ensure heating at the desired temperature, automatic temperature feedback control systems are used to monitor the microwave power input or speed of the conveyor belt in such continuous-throughput microwave heating systems. The volumetric and instantaneous nature of microwave energy, compared to conventional heat sources used in conventional heating systems, provides operators with the capability to adjust the heating rate with almost relatively instantaneous response when compared to the response needed in the case of conventional heat transfer processes. As discussed in Chapter 5, one of the major challenges in developing feedback control systems is the incompatibility of thermocouples used for monitoring temperature. Various alternative methods for the monitoring of temperature for use in microwave heating, including the use of fibre-optic sensors and infrared thermography, were discussed in Chapter 5.

Besides ensuring the supply of an appropriate level of power, another important parameter in designing continuous-throughput microwave heating systems is uniformity of heating. As discussed previously in this chapter, achieving heating uniformity is one of the biggest challenges in microwave heating. One of the advantages of using conveying systems is that the movement of the load through the cavity reduces considerably heating nonuniformity in the direction of movement. However, an important point to take into account when designing continuous-throughput microwave cavities is the possibility that the nonuniformity in heating in directions other than the direction of motion of the load itself can compound and give rise to considerable heating nonuniformity. This is especially of concern when the load being heated (e.g., concrete) has temperature-dependent dielectric properties.

6.4.3.1 Methods to minimise microwave leakage from continuous-throughput microwave heating systems

Minimising the microwave power leakage from continuous-throughput microwave heating cavities is necessary to comply with occupational safety and health regulations, avoid interference with the instrumentation in the vicinity of the heating system, and reduce wastage of energy. Among these objectives, achieving the second is perhaps the most difficult as interference with communication devices may occur in situations involving relatively low-power leakages. However, as described in Chapter 1, most commercial microwave heating systems are designed to use ISM frequencies at which the interference of the microwave heating system itself with adjacent communication and instrumentation systems is expected to be minimal.

Microwave heating systems invariably make use of metallic enclosures and gaskets to reduce leakage from doors and seams. If designed and manufactured properly, these should be sufficient to ensure the safety and health of operators (Figure 6.31). These safety measures are especially effective in multimode cavity applicators, which are usually fully enclosed with all the interfaces of the enclosure itself seamlessly welded to prevent leakages (Figure 6.32).

However, minimising microwave power leakage is usually challenging in continuous-throughput microwave heating systems as the microwave cavity has inlet and outlet ports. If not designed properly, microwave power can easily leak through these inlet and outlet ports. In addition, many microwave heating systems may require the presence of other openings for instrumentation purposes, such as openings for thermal cameras, which require a direct line of sight to the load and for optical fibre temperature sensors to pass through. Such openings contribute directly to leakage of the microwave power.

To minimise leakage of microwave power from the inlet and outlet ports of continuous-throughput microwave heating systems, these should

Figure 6.31 RF gasket of the chamber's door.

Figure 6.32 Continuous welding of the steel plates in the manufacture of multimode cavities.

incorporate attenuation ducts. Attenuation ducts are also referred to as vestibules, attenuation chocks, or attenuation tunnels. The design of such a attenuation duct depends on the size of the loads processed by the system, which also determines the minimum size of the openings required at the ports. Rectangular and cylindrical attenuation ducts are the most popular for use in continuous-throughput microwave heating systems (Figure 6.33). Openings without attenuation ducts could lead to considerable leakage of microwave energy unless the opening is much smaller than the wavelength of the operating microwave frequency.

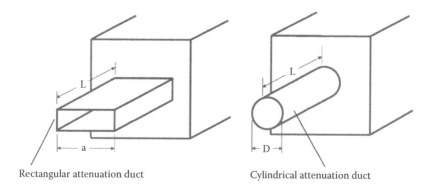

Rectangular attenuation duct   Cylindrical attenuation duct

Figure 6.33 Rectangular and circular attenuation ducts.

Table 6.2 Cutoff dimensions of rectangular and cylindrical attenuation
          ducts at selected frequencies

Frequency	Cutoff inner diameter of cylindrical attenuation duct (cm)	Cutoff width of rectangular attenuation duct (cm)
915 MHz	19.23 cm	16.30 cm
2.45 GHz	7.18 cm	6.12 cm
5.8 GHz	3.03 cm	2.58 cm

The openings of continuous-throughput microwave heating systems can be generally divided into below-cutoff and above-cutoff types. In below-cutoff openings, the wave amplitude decreases quickly with distance, and the waves do not propagate. However, in above-cutoff openings, the use of typical attenuation ducts with metallic walls is not effective unless certain techniques (introduced later) are applied.

The cutoff frequency of an attenuation duct is the frequency above which the waves can propagate in the attenuation duct itself. In rectangular attenuation ducts, the TE_{10} mode is the mode with the longest wavelength. This mode has been shown to have a cutoff frequency of $f_c = 150/a$ (in MHz), where a (in metres) is the width of the opening, as shown in Figure 6.33. In cylindrical attenuation ducts, however, the longest wavelength belongs to the TE_{11} mode, for which the cutoff frequency can be estimated as $f_c = 176/D$ (in MHz), where D (in metres) is the inner diameter of the attenuation duct. These equations can be used to estimate the cutoff dimensions of attenuation ducts at different operating frequencies. Table 6.2 lists the cutoff dimensions for rectangular and cylindrical attenuation ducts at ISM frequencies commonly used in continuous-throughput microwave-processing applications, including applications discussed in the previous chapters. If the cross-sectional dimension is larger than the cutoff dimension, substantial leakage is likely to occur regardless of the length of the attenuation duct.

However, if the dimension is smaller than the cutoff dimensions estimated, the length of the attenuation duct plays an important role in determining the amount of leakage. As a general rule, as the dimensions of the tunnel reach that of the cutoff dimensions, a longer attenuation duct is required to minimise the leakage. The required length of the attenuation duct can be estimated by calculating the attenuation constant using the following general equations for rectangular and cylindrical attenuation ducts, respectively:

$$\alpha = 0.18\sqrt{\frac{22500}{a^2} - f^2} \quad \text{(For rectangular attenuation ducts)} \qquad (6.22)$$

$$\alpha = 0.18\sqrt{\frac{31100}{D^2} - f^2} \quad \text{(For cylindrical attenuation ducts)} \qquad (6.23)$$

Here, α is the attenuation constant (in dB/m), f is the microwave frequency, a is the width of the rectangular attenuation duct, and D is the inner diameter of the cylindrical attenuation duct. In most practical applications using ISM frequencies, an attenuation level of 40 dB is satisfactory. Once the attenuation constant of the attenuation duct (dB/m) is estimated using Equations 6.22 and 6.23, the length of the attenuation duct required to reduce the attenuation level to 40 dB can be easily calculated.

As discussed, the use of attenuation ducts alone is only effective when the dimensions of the openings required are smaller than the cutoff dimensions for the operating microwave frequency. However, the dimensions of the openings required in applications such as accelerated microwave curing of precast concrete tend to be larger than the cutoff dimensions at typical microwave operating frequencies used in such applications (Table 6.2). In such situations, the attenuation ducts are of the above-cutoff duct type and require the use of additional measures to minimise power leakage from openings. When the opening is not excessively large (dimensions close to the operating wavelength), one common technique adopted is to use absorbing linings inside the attenuation ducts [17]. Another method used in such cases is to install ferritic materials inside the attenuation ducts. Most common waveguide excited modes have a strong magnetic field parallel to the walls; therefore, the use of ferritic materials, which have a high magnetic loss, in attenuation ducts is an effective method to attenuate the leaked power quickly.

However, when the required opening is considerably larger compared to the wavelength of the operating microwave frequency (multiwavelength dimensions), the waveguide modes tend to be in the TEM mode, and waveguide walls play an insignificant role in defining the boundary conditions. In such cases, covering the walls with highly lossy materials or ferritic materials tends to be not as effective. In addition, even when lossy materials are theoretically effective, their effectiveness may reduce considerably as they get heated up. Another method to reduce the leakage of microwave power from continuous-throughput microwave heating systems is the resonant choking method. Unlike the use of absorbing materials, this method relies on reflecting the emissions. The effectiveness of the resonant chocking method in reducing power leakage from above-cutoff attenuation ducts is because leakage minimisation in above-cutoff ducts is significant mainly at higher frequencies, which usually have relatively narrow frequency bands. The resonant choking method takes advantage of the narrow frequency band of such frequencies to minimise power leakage through the use of band-stop filters installed at the openings of cavity. In this method, repetitive metallic features are arranged in the attenuation duct in a way to resonate the operating frequency. A band-stop filter with such rectangular filter elements is shown in Figure 6.34 [18].

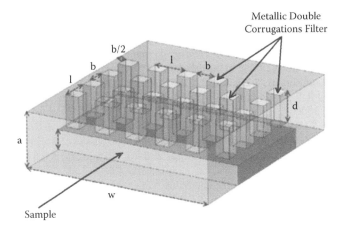

Figure 6.34 Doubly corrugated metallic filters. (From Clemente-Fernández, F.J., Monzó-Cabrera, J., et. al. Analysis of reactive and resistive open waveguide filters for use in microwave-heating applicators. *COMPEL: The International Journal for Computation and Mathematics in Electrical and Electronic Engineering*, 2011, **30**(5):1606–1615. With permission.)

6.5 SUMMARY

The effectiveness of the microwave-assisted concrete curing, concrete demolition, and concrete-recycling methods discussed in Chapters 2 to 4 depends on the efficiency of the applicators designed to apply these methods in practice. The applicators should be designed by considering the operating costs, efficiency, ease of use, and attendant safety and health issues. In this chapter, the configuration and various components of typical industrial microwave heating systems suitable for use in the applications discussed in this book were introduced. By taking into account the considerable influence of applicator characteristics on the distribution and intensity of the incident microwave field, focus was then placed on designing the microwave applicators. The basics involved in the design of three important types of applicators used in microwave-assisted concrete-processing methods as discussed in this book were presented. These applicators include open-ended waveguide applicators, multimode cavity applicators, and continuous-throughput applicators. A number of methods to improve efficiency, heating uniformity, and safety and health issues associated with the design of applicators were discussed.

REFERENCES

1. Gerling, J.F., *Waveguide Components and Configurations for Optimal Performance in Microwave Heating Systems*. Modesto, CA: Gerling Applied Engineering. Available at http://www.2450mhz.com/PDF/TechRef/Waveguide%20Components%20and%20Configurations.pdf.
2. Collin, R.E., *Foundations of Microwave Engineering*. New York: Wiley-IEEE Press, 2000.
3. Ramo, S., Whinnery, J.R., and Van Duzer, T., *Fields and Waves in Communication Electronics*. 3rd ed. New York: Wiley, 1994.
4. Ishii, T.K., *Microwave Engineering*. 2nd ed. New York: Oxford University Press, 1996.
5. Pozar, D.M., *Microwave Engineering*. New York: Wiley, 2004.
6. Mehdizadeh, M., *Microwave/RF Applicators and Probes for Material Heating, Sensing, and Plasma Generation: A Design Guide*. Norwich, NY: William Andrew, 2009.
7. Chan, T.V. and Reader, H.C., *Understanding Microwave Heating Cavities*. Boston: Artech House, 2000.
8. Akarapu, R., Li, B.Q., et al., Integrated modeling of microwave food processing and comparison with experimental results. *Journal of Microwave Power and Electromagnetic Energy*, 2004, **39**(3/4):153–165.
9. Chan, T.V., Tang, J., and Younce, F., Dimensional numerical modeling of an industrial radio frequency heating system using finite elements. *Journal of Microwave Power and Electromagnetic Energy*, 2004, **39**(2):87–105.
10. L. Fehrer et al., The industrial hephaisstos microwave system technology. *40th Annual Microwave Symposium Proceedings, IMPI*, August 9–11, 2006, Boston, Massachusetts. International Microwave Power Institute (IMPI).
11. Egorov, S.V., Eremeev, A.G., et al., Edge effects in microwave heating of conductive plates. *Journal of Physics D: Applied Physics*, 2006, **39**:3036–3041.
12. Metaxas, A.C. and Meredith, R.J., *Industrial Microwave Heating*. Stevenage, UK: Institution of Engineering and Technology, 1988.
13. Gerling Applied Engineering. Home page. http://www.2450mhz.com/ (accessed March 3, 2014).
14. Jerby, E. and Dikhtyar, V., Method and device for drilling, cutting, nailing and joining solid non-destructive materials using microwave radiation. US Patent 6114676, filed January 19, 1999 and issued September 5, 2000.
15. Lewin, L., *Advanced Theory of Waveguides*. London: Iliffe, 1951.
16. Boldor, D., Sanders, T.H., et al., A model for temperature and moisture distribution during continuous microwave drying. *Journal of Food Process Engineering*, 2005, **28**(1):68–87.
17. Meredith, R.J., *Engineers' Handbook of Industrial Microwave Heating (Power and Energy Series)*. Cambridge, UK: IEE Press, 1998.
18. Catalá-Civera, J.M., Soto, P., et al., Design parameters of multiple reactive chokes for open ports in microwave heating systems. In *Advances in Microwave and Radio Frequency Processing, 8th International Conference on Microwave and High-Frequency Heating*, M. Willert-Porada (Ed.). Berlin: Springer-Verlag, 2006, 39–47.

Index